New Geographies of the Globalized World

Globalization has, essentially, come to an end. It is, already, a victorious revolution. It has profoundly restructured the relationships between people and the world, often recreating them in a new geographical image.

This book discovers and describes these relationships of new geographies, providing a comprehensive spatial guide to the globalized world of the twenty-first century. It considers a number of timely and important themes and insights for the present and future world, exploring topics such as population trends and migration; development, the urban; transportation; religion; our endangered planet; wars, conflicts and terrorism; and disease. As such, it offers a cross-cutting synthesis of the modern world. It will be of interest to students and researchers in humanities and social sciences, including geographers, economists, political scientists and IR specialists.

Marcin Wojciech Solarz is an Associate Professor, PhD habil., Faculty of Geography and Regional Studies, University of Warsaw, Poland.

Routledge Studies in Human Geography

This series provides a forum for innovative, vibrant, and critical debate within Human Geography. Titles will reflect the wealth of research which is taking place in this diverse and ever-expanding field. Contributions will be drawn from the main sub-disciplines and from innovative areas of work which have no particular sub-disciplinary allegiances.

For a full list of titles in this series, please visit www.routledge.com/series/ SE0514

New Geographies of the Globalized World

Edited by Marcin Wojciech Solarz

Routledge
Taylor & Francis Group

LONDON AND NEW YORK

First published 2018
by Routledge
2 Park Square, Milton Park, Abingdon, Oxon OX14 4RN

and by Routledge
605 Third Avenue, New York, NY 10017

First issued in paperback 2021

Routledge is an imprint of the Taylor & Francis Group, an informa business

Translation from Polish and English consultation by Mark Znidericz.

Maps prepared by Jarosław Talacha.

Translation and map preparation financed by the Faculty of Geography
and Regional Studies and Institute of International Relations, University of
Warsaw and the Rector's Office of the University of Warsaw, Poland.

Publisher's Note
The publisher has gone to great lengths to ensure the quality of this reprint
but points out that some imperfections in the original copies may be
apparent.

British Library Cataloguing in Publication Data
A catalogue record for this book is available from the British Library

Library of Congress Cataloging in Publication Data
A catalog record for this book has been requested

ISBN 13: 978-0-367-50623-0 (pbk)
ISBN 13: 978-1-138-67641-1 (hbk)
ISBN 13: 978-1-315-56008-3 (ebk)

DOI: 10.4324/9781315560083

Typeset in Times New Roman
by Wearset Ltd, Boldon, Tyne and Wear

Contents

Figures

Tables

Boxes

Contributors

Jozsef Benedek, Professor, PhD, Faculty of Geography, Babeş-Bolyai University, Cluj-Napoca, Romania.

Voicu Bodocan, Associate Professor, PhD, Faculty of Geography, Babeş-Bolyai University, Cluj-Napoca, Romania.

Anna Dudek, PhD, Faculty of Geography and Regional Studies, University of Warsaw, Warsaw, Poland.

Ferenc Gyuris, PhD, Institute of Geography and Earth Sciences, Eötvös Loránd University (ELTE), Budapest, Hungary.

Barbara Jaczewska, PhD, Faculty of Geography and Regional Studies, University of Warsaw, Warsaw, Poland.

Attila Jancsovics, MSc, Morgan Stanley Hungary Analytics Ltd, Budapest, Hungary.

Maciej Jędrusik, Professor, PhD habil., Faculty of Geography and Regional Studies, University of Warsaw, Warsaw, Poland.

Imre Keserü, PhD, Mobility, Logistics and Automotive Technology Research Centre (MOBI), Vrije Universiteit Brussel, Brussels, Belgium.

Izabella Łęcka, PhD habil., Faculty of Geography and Regional Studies, University of Warsaw, Warsaw, Poland.

Cathy Macharis, PhD, Mobility, Logistics and Automotive Technology Research Centre (MOBI), Vrije Universiteit Brussel, Brussels, Belgium.

Marek Madej, PhD, Faculty of Political Science and International Studies, University of Warsaw, Warsaw, Poland.

Jerzy Makowski, Professor, PhD habil., Faculty of Geography and Regional Studies, University of Warsaw, Warsaw, Poland.

Joanna Miętkiewska-Brynda, MA, Faculty of Geography and Regional Studies, University of Warsaw, Warsaw, Poland.

Balázs Németh, MSc, Department of Planning, Aalborg University (AAU-CPH), Copenhagen, Denmark.

Vilmos Oszter, MSc, KTI Institute for Transport Sciences, Budapest, Hungary.

Raularian Rusu, Associate Professor, PhD, Faculty of Geography, Babeş-Bolyai University, Cluj-Napoca, Romania.

Anna M. Solarz, PhD, Faculty of Political Science and International Studies, University of Warsaw, Warsaw, Poland.

Marcin Wojciech Solarz, Associate Professor, PhD habil., Faculty of Geography and Regional Studies, University of Warsaw, Warsaw, Poland.

Gábor Szalkai, PhD, Institute of Geography and Earth Sciences, Eötvös Loránd University (ELTE), Budapest, Hungary.

Tomasz Wites, PhD, Faculty of Geography and Regional Studies, University of Warsaw, Warsaw, Poland.

Małgorzata Wojtaszczyk, PhD, Faculty of Geography and Regional Studies, University of Warsaw, Warsaw, Poland.

Acknowledgements

This book could not have been completed without the kindness and help of many people. We wish to thank the former dean of the Faculty of Geography and Regional Studies at the University of Warsaw – Professor Andrzej Lisowski, and a former pro-vice-chancellor of the University – Professor Alojzy Nowak, both of whom supported the project with the significant financial means necessary to translate and perform the initial language editing of the book, and supply it with maps and other figures. Especial thanks must go to our translator Mark Znider-icz, who has patiently persevered with material which ranges widely in content and style. I am very grateful for the contribution made by Jarosław Talacha, who put our cartographic ideas into the form of professional maps. Of course, as usual, responsibility for all shortcomings and errors found within these pages belongs only to the authors.

Introduction

Marcin Wojciech Solarz

Globalization has, essentially, come to an end. In fact, we are already living in a globalized world, even though it still has different densities in different parts of the oecumene, and inequalities in the level of globalization are still clearly evident. Our planet can be compared to an enclosed glass sphere, which we have filled with our civilization, as if with a gas. We are still unable to equalize the developmental pressure throughout the sphere at the level of the most advanced societies – and perhaps this goal cannot be achieved with the current structure and organization of the international community, and given the domination of various individual and collective egoisms over global solidarity – yet the slogan 'Yes We Can' in the context of this task does not sound quite as alien and abstract today as it once might have done.

On the eve of the Age of Discovery, which launched the Revolution of Globalization, the principles by which the world was organized were those of isolation and spatial separation. The great land masses – the Old World encompassing Africa, Asia and Europe, and the New World of the two Americas, Australia and Antarctica, as well as the many islands scattered across the great oceans, existed as if under opaque glass domes. And within these 'domed' macroworlds, local communities lived isolated from one another in a mosaic of autonomous microworlds, also domed, and so with limited contact with the world outside the horizon of everyday life. The Revolution of Globalization shattered the opaque domes and brought with it a new principle by which the world was to be organized – the principle of interdependence. Nevertheless, until the outbreak of the Industrial Revolution, travel (and, therefore, communication) was relatively slow (and, therefore, still limited in extent). The time needed to travel 1 kilometre on land ranged between twelve minutes (for an average walker) and fifty-one seconds (the approximate time for a galloping horse). Human and animal endurance was a significant limitation. There was no question of constant motion at high speed. Thus, the above 'price' of a kilometre was rather for the first one than the next, and the longer the time travelled the higher the 'price' became. In the mid-nineteenth century, it took a stagecoach no less than seventy hours to travel between Paris and Marseille, so the average 'price' of a kilometre on this route was about 5.5 minutes. By contrast, today a plane covers the same distance in less than 1.5 hours, so the 'price' of a kilometre has dropped to about six seconds. The Industrial Revolution definitively and irrevocably ripped the world from the anchor of old relationships and propelled it forward beyond the horizons of past possibilities. It transported communities with fixed horizons into a world of moving horizons. It dramatically deepened, accelerated and strengthened the Revolution of Globalization. Thus, in our glass sphere today there are over seven times more people than in the early nineteenth century, who travel faster, more often and further. The space around us has clearly become denser. Today, we often use the words 'international' (community, relations, travel, news, law), 'global' (crises, challenges, organizations, conferences, celebrities, pandemics, terrorism) and 'world' (markets, wars, leaders, tournaments). This is because globalization is, already, a victorious revolution. It has profoundly restructured the relationships between people and the world, often recreating them in a new geographical image. In this book, prepared in and focused on the globalized world of the twenty-first century, we attempt to discover and describe these relationships – new geographies, focusing on a number of timely and geographically important global themes as well as significant insights and ideas for the present and future world: population trends and migration (Chapter 1); development (Chapters 2 and 3); cities (Chapter 4); transportation (Chapter 5); religion (Chapter 6); our endangered planet (Chapter 7); wars, conflicts and terrorism (Chapter 8); and disease (Chapter 9).

1   Geographies of world population

Demographic trends in the contemporary world

Barbara Jaczewska, Tomasz Wites,
Marcin Wojciech Solarz, Maciej Jędrusik and
Małgorzata Wojtaszczyk

A measure of the speed of demographic change is the time it takes for a population to double. It is estimated that during the New Stone Age (7000 BCE) about 10 million people lived on Earth. A further 2,500 years were to pass before the world's population reached 20 million. It doubled again (from 20 to 40 million) 2,000 years later, and once more (from 40 to 80 million) after another 1,500 years. By the first century AD, some 250 million people lived on Earth. From the end of the fifteenth century population, numbers began to rise at a greater pace, but this was tempered by the death rate, which remained high. It was only in later centuries, with the advancement of medicine, that demographic change gained significant momentum. Just 330 years were needed for the world's population to double (from 500 million *c.*1490 to 1 billion *c.*1800).

Milestones in the demographic development of the world's population can be precisely, albeit rather arbitrarily, identified. Thus, humanity reached 1 billion in 1804, the year in which Napoleon Bonaparte proclaimed himself 'Emperor of the French'. The threshold of the second billion was crossed in 1927, the year of Charles Lindbergh's solo flight across the Atlantic, and the third in 1960, which was proclaimed the 'Year of Africa'. Humanity reached its fourth billion in 1974, when Richard Nixon resigned from office, and its fifth in 1987, when Ronald Reagan in his Brandenburg Gate speech called on Mikhail Gorbachev to tear down the Berlin Wall. Six billion was exceeded in 1999, when the first countries of the former Eastern bloc were accepted into NATO, and the 7-billion threshold was crossed in the year of the tidal wave of the 'Arab Spring', the death of Osama bin Laden and the wedding of Prince William and Kate Middleton (2011). No fewer than six of these seven milestones have been passed in the lifetimes of people still (as of May 2017) present among us, such as Queen Elizabeth II, Alan Greenspan and George H.W. Bush (Solarz and Wojtaszczyk 2015: 802–803).

Drawing the attention of world public opinion to the importance of these dynamic population processes and the problems they cause is one of the tasks of the Governing Council of the United Nations Development Programme (UNDP). In 1989, this body proclaimed 11 July as World Population Day, which refers back to 11 July 1987, the approximate date when the world's population passed the 5 billion mark. The difficulties involved in precise population measurement

First published 2018
by Routledge
2 Park Square, Milton Park, Abingdon, Oxon OX14 4RN

and by Routledge
605 Third Avenue, New York, NY 10017

First issued in paperback 2021

Routledge is an imprint of the Taylor & Francis Group, an informa business

Translation from Polish and English consultation by Mark Znidericz.

Maps prepared by Jarosław Talacha.

Translation and map preparation financed by the Faculty of Geography
and Regional Studies and Institute of International Relations, University of
Warsaw and the Rector's Office of the University of Warsaw, Poland.

Publisher's Note
The publisher has gone to great lengths to ensure the quality of this reprint
but points out that some imperfections in the original copies may be
apparent.

British Library Cataloguing in Publication Data
A catalogue record for this book is available from the British Library

Library of Congress Cataloging in Publication Data
A catalog record for this book has been requested

ISBN 13: 978-0-367-50623-0 (pbk)
ISBN 13: 978-1-138-67641-1 (hbk)
ISBN 13: 978-1-315-56008-3 (ebk)

DOI: 10.4324/9781315560083

Typeset in Times New Roman
by Wearset Ltd, Boldon, Tyne and Wear

of rapid population growth, unprecedented in history. A key factor in global development according to Simon is our human capacity to create new ideas and expand our body of knowledge. Based on a detailed analysis of indicators such as mortality, illiteracy, agricultural production per capita and pollution, he concludes that the more people who can be trained to assist in solving the Earth's economic and environmental problems, the sooner we will witness dynamic economic development and the richer the legacy that we will leave to future generations will be.

Population distribution within the boundaries of the oecumene (7.4 billion in 2016) is uneven. The territories constituting oecumene, suboecumene and anoecumene are subject to numerous changes. The boundaries between inhabited and uninhabited areas are difficult to define with precision. Within the oecumene, there are often areas which as a result of administrative decisions are inhabited only periodically or are completely uninhabited (e.g. areas under legal protection such as nature reserves and military bases).

The transformation of anoecumene into suboecumene or oecumene occurs in areas at varying levels of economic development. The large population growth currently observable in Africa has contributed to a change in population concentration and land development. Areas of anoecumene in the equatorial zone have become suboecumene as fragments of rainforest have been converted into farmland. The conversion of uninhabited terrain into periodically or continuously inhabited land also occurs in the mountain regions of economically highly developed countries. Some current suboecumene areas until recently were anoecumene (e.g. high-altitude resorts with both permanent and seasonal inhabitants, which were established due to the development of tourism in the twentieth century).

Oecumene and suboecumene become anoecumene much less frequently. There are several reasons why this is so, in particular the finite space of inhabited areas and the constantly increasing population of the Earth. Regions experiencing depopulation exemplify the shift from oecumene to suboecumene or anoecumene.

Changes in population: continents and countries

Beginning our analysis of population change on a continental scale, it is worth noting that between 1800 and 2000 the population of Europe increased only fourfold, while that of Africa, the site of the oldest hominid remains, saw a ninefold increase and North America's population grew fiftyfold. The least populous continent in the early twenty-first century (just 0.5 per cent of total global population) remains Oceania (Australia and the Pacific islands) (Worldometers 2017), whereas Asia has the world's largest population, thanks to its environmental conditions which are more favourable than in Africa.

At the beginning of the twenty-first century, Asia is home to some 60 per cent of total world population, followed by Africa, though the latter has the world's fastest rate of population growth. Next in terms of population size is Europe, which for many centuries was in second place (after Asia). At the beginning of the twentieth century, Europe was home to as many as 25 per cent of the world's inhabitants, but 100 years later, this proportion has fallen to less than 12 per cent. According to demographic forecasts, in 2025, Europe is expected to combine a low birth rate with clearly positive net migration and it will be inhabited by only 6 per cent of the planet's population. This sharp decline in the Old Continent's share in world population is not due to falling numbers in Europe, but instead by the rapid population growth on other continents.

When we analyse population change on a country scale, we should remember that in addition to natural and migration movements, administrative decisions can also play a role in the rise and fall of population numbers (as in the case of the reduction or enlargement of a country's territory). As recently as 1950, among the ten most populous countries in the world, there were four countries wholly or partly situated in Europe – the Soviet Union (the majority of whose population was concentrated west of the Urals), the Federal Republic of Germany, the United Kingdom and Italy. In 2016, however, only one European country can be found in the world's top ten in terms of population: Russia (most of whose citizens are located in its European part). This merely confirms the fact that contemporary rapid population increases are mainly occurring in non-European countries.

Only five countries (or just 2.6 per cent of the world's total number of countries) – China, India, USA, Indonesia and Brazil – have more than 200 million inhabitants (2016), but their populations taken together represent 47 per cent of the global total. Only thirteen countries (6.6 per cent) have more than 100 million inhabitants, but they account for more than 60 per cent of the planet's population. Not counting Russia, seven of these are located in Asia, three in the Americas and two in Africa. In Asia, the race is on for the country with the highest population there and also globally. In 1950, the population of China was 544 million and that of India was 371 million, whereas although China still held first place in 2010, demographic forecasts indicate the likelihood that in 2022 India will take the lead.

An extremely important characteristic of the world's population since the second half of the twentieth century is a clear difference in the demographic processes taking place in more economically developed countries compared to their less developed counterparts. As recently as the mid-twentieth century, the latter accounted for 67 per cent of global population, whereas by 1980, this figure had risen to 75 per cent and in 2015, it was 85 per cent (Worldometers 2017).

Demographic forecasts

By way of introduction to the subject of this section, it is worth considering a representative opinion concerning the validity of preparing and attaching weight to population projections. Secomski notes that the publication of demographic forecasts, and the interest which they widely attract, has led to the formulation of new principles of population policy (1978: 16–18). The majority of forecasts concerning the development of the world's population have not stood the test of time. This does not mean, however, that they have had no impact on shaping the world in social, economic, political and even moral terms. Notwithstanding the widespread reliance on demographic projections in development plans and programmes, Secomski argues that we should not overestimate their role (1978: 18).

Before key aspects of demographic forecasts relating to the modern world are presented here, we will reflect, albeit briefly, on past projections whose accuracy we are now able to verify. This will give the reader some perspective from which to assess the value of today's forecasts. In 1798, the economist Thomas R. Malthus published an important work titled *An Essay on the Principle of Population*, which forecast a sharp rise in population and predicted that the effects of uncontrolled demographic expansion would be dire. Although others had issued such warnings much earlier, Malthus was the first to formulate a comprehensive theory ('the Iron Law of Population'), based on the assumption that population growth progresses geometrically, whereas food resources grow arithmetically. From this, the English scholar drew the conclusion that there was no room at the 'table of nature' (Malthus 1798: 4–5) for an ever-increasing number of people.

Although Malthus' theory contained certain flaws, it had a profound impact on public opinion at the time of publication, and continues to exert influence despite the passage of more than 200 years. Many contemporary analyses of demographic trends draw on Malthusian theory, which has its origins in a debate between its author and his father. The latter was an advocate of the theory propounded by William Godwin, according to which poverty in the world is the consequence of defective social institutions, and so could be effectively limited if there were a more equitable distribution of national income. Malthus junior did not share this view, rather, he believed that the source of poverty was excessive population growth (1798: 55–67). The theses of Malthus were later contested by the American economist H.C. Carey, who argued that population adapts to conditions, and as a result, effective overpopulation does not occur (1867: 10). When the population increases, an association factor becomes operative, manifesting itself in a better division of labour being possible, leading to a disproportionate increase in productivity, and thus overpopulation is prevented. The larger a population is, the greater the diversity of talents and the more effective the division of labour.

An example of a scholarly publication based on demographic forecasts and models deduced from them is *The Limits to Growth* (Meadows, Meadows, Randers and Behrens 1972). The analysis it contains, carried out for the Club of Rome, weighs the future of humanity in light of the increase in the number of the Earth's inhabitants and the claimed dwindling of natural resources. The forecasts set out in the book proved correct but only to a limited degree, and the majority of them, concerning the production of food, were contradicted by estimates and projections later prepared by other institutions, including the International Food Policy Research Institute (Penning de Vries, Van Keulen, Rabbinage and Luyten 1995).

Taking into account various population change scenarios, the United Nations estimates that in 2050 between 7.7 and 10.6 million people will live on the Earth, with the most likely figure reckoned at 9.1 billion (United Nations 2005: 16). In 2015, the United Nations prepared population forecasts up to the year 2100. Two variants – high and medium – predict an increase in world population to 16.5 billion and 11.2 billion respectively by 2100. Only the low variant foresees a population decrease to below today's values. Given the large number of forecasts, often based on contradictory assumptions, we can conclude that some of them definitely will not prove correct. It should be noted, however, that most projections do foresee a further strengthening of disparities in population numbers between different regions of the world (Kalaska and Wites 2015: 1819–1822). This is due to rapid population growth in the countries of the global South and the gathering pace of depopulation processes in substantial parts of the rest of the world.

In each of the three variants (high, medium and low growth), Africa rivals Asia in terms of population numbers. In 2100, according to the high growth variant, Asia will be the world's most populous continent, whereas in the medium variant the two continents are very closely matched, and in the low variant Africa is the predicted world leader. Among the African countries, the highest population growth up to 2100 will most likely be seen in Nigeria, the Democratic Republic of the Congo, Ethiopia and Uganda, and India, China, Pakistan and Bangladesh will lead in Asia. In the high variant, all the world's continents will have higher populations in 2100, and in the medium variant only Europe will have seen a numerical reduction, whereas the low variant foresees that Europe will be joined by Asia and the entire western hemisphere.

Demographic transition

Changes in world population can be described using the concept of demographic transition developed in the 1930s by the Institute of Population Research at Princeton University in the United States (Caldwell 2006: 301). According to this concept, changes in population numbers in a given area are the result of natural movement, whose constituents are marriage, divorce, birth, death and migration. Demographic transition is the process of change in population reproduction associated with the modernization of societies (Notestein 1945: 39). In a narrower sense, this means a significant reduction in birth and death rates, accompanied at first by a systematic increase in the rate of population growth, then its steady decline (the birth rate decreases faster than the death rate). In a broader sense, demographic transition means a qualitative transformation whereby traditional population reproduction is replaced by modern reproduction.

On a theoretical diagram of the demographic cycle (Figure 1.1), we can trace the course of population change in different countries. It should be remembered that demographic transition in particular societies can last from a few dozen to over 100 years, and people's reproductive behaviour is based on the human ability to adapt to the surrounding environment. The spread of demographic

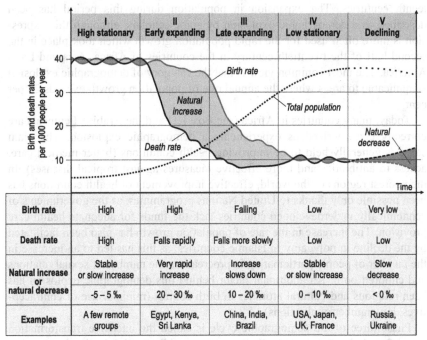

	I High stationary	II Early expanding	III Late expanding	IV Low stationary	V Declining
Birth rate	High	High	Falling	Low	Very low
Death rate	High	Falls rapidly	Falls more slowly	Low	Low
Natural increase or natural decrease	Stable or slow increase	Very rapid increase	Increase slows down	Stable or slow increase	Slow decrease
	-5 – 5 ‰	20 – 30 ‰	10 – 20 ‰	0 – 10 ‰	< 0 ‰
Examples	A few remote groups	Egypt, Kenya, Sri Lanka	China, India, Brazil	USA, Japan, UK, France	Russia, Ukraine

Source: Own elaboration based on Worldometers (2017)

Figure 1.1 The demographic transition model.

modernization takes the form of diffusion, so it may be presumed that the overall pattern of population change taking place first in European countries will be repeated in a similar form in other countries around the world.

In phase one of the demographic cycle, there is low population growth arising from very high rates of reproduction combined with very high mortality, especially among infants and women after childbirth. This phase is characteristic of traditional societies, in what is known as the pre-industrial era. The close of this phase in Western Europe took place in the mid-eighteenth century. At the beginning of the twenty-first century, this type of natural reproduction occurs only in isolated communities, for example, in the Amazon, New Guinea and some island regions (such as the Andaman Islands).

Phase two of the demographic cycle is called the early transformation stage. It is marked by a sharp decline in mortality while reproduction remains at a high level, which in large measure is due to medical advancements, increased access to medicines and improved hygiene and sanitary conditions. Significant population growth in Western Europe occurred in the eighteenth century with the onset of the industrial revolution in Great Britain, although it was not until the late nineteenth century that this increase became exponential (a 500-million jump in population between 1820 and 1890). In the remainder of Europe, phase two of significant population growth took place at the turn of the eighteenth and nine-teenth centuries. The expansion in population during this period has been described as a demographic explosion by some scholars, although this expres-sion is more often used for the rapid population growth which took place in the second half of the twentieth century in the countries of Africa, Asia and Latin America. The most common view is that the category of demographic explosion is applicable to areas where the annual rate of population growth exceeds 20 per cent.

Today, many countries in Africa (see Box 1.1) and the Arabian Peninsula are correctly characterized as experiencing a demographic explosion. The main cause is generally held to be improving health conditions (better medical care, access to antibiotics and more effective measures against tropical diseases). In the poorest regions of the world, effective improvement of health conditions has been possible only thanks to United Nations programmes, as the governments of economically underdeveloped countries lack the funds for adequate health care provision. The increase in the rate of population growth has also been facilitated by the decline in polygamy in Islamic countries, as this has led to an increase in the number of people participating in procreation, the number of sexual relations and children born. An additional factor behind this demographic explosion has been religious and cultural attitudes to birth control (minimal use of contracep-tives, low number of abortions).

Phase three of the demographic cycle is called the intensive transformation stage. Population numbers continue to grow rapidly, as the decline in mortality is accompanied by large, albeit now diminishing, birth rates. At this point, we see the conscious regulation of the number of births. Western Europe passed through this phase in the first half of the nineteenth century, while the develop-ing countries of Asia and Latin America are experiencing it in the first two decades of the twenty-first century. Phase four is called the late transformation stage. It is marked by a slowing decline and then stabilization in mortality, and a dramatic reduction in the number of births. Population growth shifts from explo-sive to implosive. This situation is the result of many factors, including women entering employment. Phase four is characteristic of economically developed countries, for example, the USA.

Phase five of the demographic cycle is called the end of transition stage or the modernized society phase. Low mortality combined with low fertility results in population numbers growing only slightly, or even falling. The phenomenon where the number of deaths exceeds the number of births is described as demo-graphic scissors. If the population loss is not compensated by positive net migra-tion, then depopulation occurs. In general, negative quantitative changes are accompanied by adverse demographic and social changes in the spatial dimen-sion. Depopulation often occurs in areas with a peripheral location (Wites 2006: 71–79), for example, certain regions of Sweden, Denmark and Canada, and it has also affected Eastern European countries which entered a period of deep crisis following the collapse of the USSR, such as Georgia, Ukraine and Russia. Areas characterized by population loss are generally single-function (dominated

Box 1.1 Demographic explosion: the example of Ethiopia

There are countries in Africa where traces of prehistoric man have been found which are older than in Ethiopia, for example, *Orrorin tugenensis* in Kenya is dated at 5.7 million years old (Pickford and Senut 2001: 22). Certain countries are also currently experiencing a larger demographic boom than Ethiopia, for example, Niger with a population growth rate in 2015 of 41 per cent (Worldometers n.d.a). Ethiopia is, nevertheless, a special country which is worth looking at more closely in order to understand the complexity of the transformations which take place in the early stage of demographic transition, known as the demographic explosion phase. Ethiopia is the only country in Africa outside Egypt with such a rich history, symbolically opened by Lucy – *Australopithecus afarensis* who lived 3.2 million years ago. Discovered in 1974, Lucy was put on public display in the National Museum in the Ethiopian capital, Addis Ababa, and named in Amharic *Dinkenesh*, 'beautiful'. The same can be said, from a modelling perspective, of the demographic development of Ethiopia, which is described as an island of black Christianity surrounded by a sea of Islam. The most populous ethnic groups (the Oromo and Amhara peoples) as well as the smaller tribes (including the Mursi, Karo and Dora) have distinct demographics. Despite the limitations of the natural environment and problems in obtaining sufficient food, Ethiopia's population continues to increase. Population growth has occurred here since ancient times, although its pace, causes and effects have been highly varied spatially. Due to its scale, the increase in population seen in Ethiopia today is referred to as a demographic explosion (in 2015, the rate of population growth in Ethiopia was 25.7 per cent). For several decades, based on the difference between the number of deaths and births, the population of Ethiopia has increased by between 1.5 and 2 million people annually. Thus, in a country where in 1950 there were only 18.1 million people, in 2015, the population reached 99.4 million. Demographic projections indicate further dynamic growth – to 138.3 million in 2030 and then 188 million in 2050 (Worldometers n.d.b). If these projections prove accurate, this will mean that, in the period 1950–2050, the population of the country will have increased tenfold.

Excessive population growth is frequently blamed for Ethiopia's significant difficulties in overcoming underdevelopment and poverty. The demographic development of Ethiopia is becoming a challenge, but not necessarily one meriting only pessimistic predictions. The Independent Inquiry Report into Population and Development (IIRPD) affirmed the revisionist view that over time a positive relationship becomes visible between population growth and sustainable development. In the report's summary, Amartya Sen, winner of the 1998 Nobel Prize in economics, argued that famines in developing countries are not caused by lack of food, but are the result of incompetent action by government institutions (2000: 3–4).

by one activity such as farming or industry). The causes of depopulation are complex, although political change and economic instability are often key factors. The scale of depopulation is determined by institutional, organizational and technological factors. Equally important are random factors expressed in the specific features of the distribution of particular communities (Wites 2007: 133–172). In countries experiencing depopulation processes (see Box 1.2), fertility decreases to below the intergenerational replacement level, which is characteristic of what has been called the 'second demographic transition' (Van de Kaa 2004: 4–10). This concept denotes societies in which fertility has stabilized at a level where the birth rate is lower than the mortality rate.

Challenges of global migration

Already in the 1930s, the outstanding Polish geographer, Stanisław Pawłowski, vividly drew attention to the close interrelation and conflicting nature of the phenomena of population growth, as well as the mobility of humanity on a planetary scale:

> If the Earth is likened to a large multi-storey building, then the world's various countries and states would, in this comparison, be its myriad apartment dwellings. Societies and countries are tenants on the Earth. Flooding over its dry surface, humanity is not, in a broad time perspective, a permanent resident of any piece of the globe. Nations and states undergo change and ceaselessly move their estates and territories. So it has always been and will be for as long as the Earth is inhabited by such a restless element as humanity. The slow concentration of mankind on Earth makes these changes ever more difficult. The interests of states and nations collide with ever greater frequency. The Earth is becoming increasingly cramped and uncomfortable.
>
> (Pawłowski 1938: 199–200)

In the modern world, few phenomena affect all areas of life as greatly as migration. Migrant flows permanently change the social structures of both the host and originating countries and pose new challenges for them. The significance of contemporary migrations is not due to their statistical size. For unlike humankind's earlier migratory explorations, today's people movements do not melt into empty expanses ripe for settlement, but are concentrated in countries which are highly saturated in terms of population, economy and civilization. In addition (and more importantly, as it seems), the significance of contemporary migration flows arises from the complexity of the effects which potentially accompany them.

Current migration processes are extremely diverse, multidimensional and difficult to describe. We are seeing both a continuation of established migration processes (which are taking on new forms), and new (previously unknown) types of population movements resulting from economic and cultural changes, as well as military conflicts. Due to the enormous diversity of contemporary migration

Box 1.2 Demographic implosion: the example of Russia

Russia is the largest area on Earth (17.1 million km^2) in which depopulation processes are taking place. It should be emphasized that there are no military conflicts
in the area of the country subject to depopulation. Population growth last occurred
Russia in 1991, the year of the dissolution of the Soviet Union (an increase in that
year of 104,000). Since 1992, there has been a systematic decline in the population
(in 1992, a loss of 219,000; in 1997, a loss of 756,000; and in 2001, as many as
943,000 people). According to demographic forecasts, 135.6 million of people will
be living in Russia in 2030, and 128.6 million in 2050 (Worldometers n.d.c). The
progressive depopulation of Russia combined with demographic growth in the
world's most populous countries means that Russia's weight in the global population is systematically declining – in 1991, Russia ranked sixth in this measure,
whereas according to forecasts, it will occupy fifteenth place in 2050.

The process of depopulation in Russia is occurring due to the simultaneous
cumulation of two processes: a marked decrease in the number of births and a rise
in the number of deaths. In the immediate post-Soviet period, in some territorial
units of Russia, infant mortality rates reached record values (e.g. in the Evenki
Autonomous Region of Siberia in 1997, the figure was 52.8 per cent, while the
average for Russia was 19.6 per cent). The very high infant mortality in Russia's
peripheral northern regions is linked to exogenous factors – unfavourable natural
conditions, poor hygiene, poor nutrition and birth trauma (Wites 2007: 133–172).

Some scholars believe that the socio-economic turmoil which followed the collapse of the Soviet Union only sped up the process of the second demographic
transition in the former Soviet republics, including Russia (Zakharov 2008:
907–972). It is very likely that had there been no post-Soviet crisis, low fertility
would have occurred in the area only 10–15 years later. There is no doubt,
however, that the economic hardship after 1991 had a heavy impact on young
couples who as a result were less likely to have children, since without parental
help, they would not have been able to function normally. During a period of systemic change, the inability to provide decent living conditions for children leads to
some couples postponing the decision to start a family, or to relationship strains in
families with children, resulting in an increasing divorce rate and a diminishing
birth rate, which distorts the age structure. Certain of the factors in the depopulation of Russia are 'regulators', which influence the size and intensity of the
process. Among these are various social pathologies, some new and some already
present before 1991 but intensified since, which contribute to unfavourable demographic change. Such pathological phenomena include: alcoholism, drug addiction,
murder and suicide. Vodka is the direct or indirect cause of 30 per cent of male
deaths in Russia. The number of alcohol addicts in Russia is estimated at 10
million (Wites 2007). According to the World Health Organization (WHO), in
2010, the consumption of alcohol in Russia measured in litres of pure alcohol per
head was 15.1 (23.9 for men and 7.8 for women). In this respect, Russia ranks
fourth in the world, after Belarus (17.5 litres per year), Moldova and Lithuania.

flows, our description of them here is limited to the five most important tendencies which are essential to their characterization.

The first of these is the *extensive spread (globalization) of migration*. The mass population movements we see today have been caused by the rapid integration of countries on a global scale and in turn, they have become one of the most important factors in global change. Migration is not an isolated process, and almost always population flows are prompted by flows of capital and goods. Cultural exchange, faster means of transport and the proliferation of electronic media also contribute. The globalization of the world economy and the continually changing poles of economic activity have brought in their wake dynamic flows of people into new centres of development, but also the collapse of traditional economic systems (agriculture, crafts and trade), thus forcing further migrations. In decisions to migrate, individual choice is increasingly giving way to collective movements of an almost spontaneous nature. In this context, great importance has been gained by a variety of migration networks based on family and ethnic ties, and also professional connections as migration has become a branch of international business.

Stephen Castles, Hein de Haas and Mark J. Miller (2014) point to the existence and increase in number of various groups who earn a living from organizing migration flows, which they have dubbed 'the migration industry'. Among these groups are: travel agents, employee recruiters, intermediaries, translators, real estate agents and lawyers, but also people smugglers and document counterfeiters. The development of the migration industry is an inseparable aspect of social networks and transnational connections, but it is also a driving force of migration streams.

An important region of the world in relation to contemporary economic migration (legal and illegal) is the Persian Gulf, and it is here that we can trace the rise of a specific migration industry, the precursor of today's global network. The oil crisis of 1973 was of great importance for the development of migration in the region. The sudden increase in oil prices generated financial resources thanks to which the oil exporting countries began to carry out costly infrastructure projects. The boom in construction required the employment of thousands of foreign workers, generating significant population flows, mainly of temporary workers. The parallel expansion of the service sector contributed to a further increase in the number of migrants. *Kafala*, or the traditional system of sponsorship, which lies at the root of the modern migration industry in the region, was based on agreements concluded between the local emir and foreign oil companies. Under the agreement, a *kafil* (sponsor and special 'migration agent') sought trustworthy people (usually Bedouins) to work in the oil fields. As the oil industry expanded, workers began to be recruited from abroad and the *kafil* served as an official intermediary between the foreigner and the administration, and between the local authorities and the local community.

From the mid-1960s to the mid-1970s in the Gulf countries, most migrants were Arabs (mainly Egyptians, Yemenis, Palestinians, Jordanians, Lebanese and Sudanese). From the 1990s, however, the proportion of non-Arab workers began to grow, mainly from South and South-East Asia (India predominantly, but also Bangladesh, Pakistan, Indonesia and the Philippines). The growth in migration from Asia (and increasing feminization of migration to the region) was accompanied by a deterioration in working conditions.

The main countries of destination in the region for international migrants in 2015 were Saudi Arabia, the United Arab Emirates, Jordan, Kuwait and Lebanon. The number of migrants in Saudi Arabia was estimated at over 10 million in a country of 27 million inhabitants in total (the fourth largest migrant population in the world), while the UAE had an estimated 8 million migrants (with 9.3 million inhabitants in total). In percentage terms, over 86 per cent of the population in the UAE were foreign; in Qatar: 73 per cent and Kuwait: 70 per cent. It is worth noting that the proportions of foreigners to total population in the countries of the region are among the highest (and in some cases, actually the highest) in the world.

Temporary, blue-collar migration from Asia can be described as 'incomplete migration' (Okólski 2012), that is, characterized by short-term moves between countries (often circular) by unskilled workers, often in what are called '3D' jobs (dull, dirty and dangerous).

The second characteristic of contemporary population movements is their *mass scale*. It should be noted in this context that the exact magnitude of world-wide migration is unknowable. In statistical analyses, we most commonly use estimates from, for example, the Population Division of the United Nations Department of Economic and Social Affairs (UNDESA), the International Migration Organization and the World Bank. However, since not all migratory movements are captured in the international statistics, we may safely assume that the scale of the flows is, in fact, much higher than recorded.

UNDESA data indicates that the last fifteen years have seen a steady and significant increase in the number of international migrants: from 173 million in 2000 and 222 million in 2010 to 244 million in 2015 (UNDESA 2016). The number of migrants is rising faster than the increase in the number of the Earth's inhabitants. As a result, the proportion accounted for by migrants rose to 3.3 per cent of world population in 2015, compared to 2.8 per cent in 2000. In the period 2000–2015, the number of migrants increased in 167 countries or areas worldwide, and in nineteen of these (including Spain, Italy, Thailand and the United Arab Emirates) migrant stock grew at a record pace, reaching an annual growth of over 6 per cent. Today, net migration has a significant impact on population numbers – it can counteract the negative trends of population ageing and decline in host countries, but also deepen these processes in areas affected by emigration.

The mass scale of migration can be considered from the perspective of the degree of concentration of migrants in different regions of the world. Almost two-thirds of all international migrants live in Europe (76 million), Asia (75 million) and North America (54 million). Since the 1990s, these areas have also seen the largest increase in the number of migrants (Table 1.1). Over 50 per cent of the total number of international migrants are concentrated in only 10 countries. In 2015, the largest number of migrants lived in the United States

Table 1.1 The number of international migrants in the period 1990–2015 according to UNDESA estimates (million)

	1990	2000	2005	2010	2015
World	152.6	172.7	191.3	221.7	243.7
Africa	15.7	14.8	15.2	16.8	20.6
Asia	48.1	49.3	53.4	65.9	75.1
Europe	49.2	56.3	64.1	72.4	76.1
Latin America and the Caribbean	7.2	6.7	7.2	8.2	9.2
Northern America	27.6	40.4	45.4	51.2	54.5
Oceania	4.7	5.4	6.0	7.1	8.1

Source: own elaboration based on UNDESA (2016).

(47 million, which accounted for 19 per cent of the world's 'migrant resources'). Germany and Russia were equal second (12 million in each country), followed by Saudi Arabia (10 million) (UNDESA 2016).

The third characteristic of contemporary migration is the *differentiation of migration*. In the last half-century, there has been a significant evolution in the motives for migration and the socio-professional structure of migrant populations. Until the 1970s, researchers were basically agreed that the subject of their research was voluntary migration, primarily with economic motives. Soon after, however, came the mass migration of refugees due to conflicts and natural disasters, which assumed dimensions equalling economic migration. Today, countries are experiencing both voluntary and forced migrations, involving white- and blue-collar workers, men and women, young and old. We are, therefore, seeing the merging of various streams of migration. Steven Vertovec (2007) uses the term 'superdiversity' to denote the contemporary level and type of diversity in societies influenced by immigration. Superdiversity denotes the 'game' which takes place between new, small and scattered groups of multiple origin, who have transnational connections and diverse socio-economic and legal statuses. An example, which well illustrates superdiversity, is contemporary London and its metropolitan area (Jaczewska 2011).

Modern migrants seem also to be adopting 'liquid migration strategies' (Black, Engberson, Okólski and Panţîru 2010). Attention to the liquidity of migration as one of the characteristics of contemporary population movements draws on Zygmund Bauman's perspective on 'liquid modernity' (2000), which highlights 'the incessant mobility of the (post-)modern era with its metaphorical figures of the tourist and the vagabond (and the migrant, the refugee, the pilgrim, etc.) arranged in a "kinetic hierarchy" of voluntary and forced mobility and immobility' (King 2012). Today, migration destinations are subject to very rapid change, and active choices are made to move to where opportunities are available at a given moment. Migrants are reacting very quickly to changes, especially those who are political and economic migrants.

Another process, which is currently visible, is the 'spillover of migration' into ever larger areas both on a global scale and at the level of regions and countries.

Migrants are no longer choosing as destinations only the largest cities and central areas, but they are travelling to places previously untouched by migration inflows (e.g. small- and medium-sized towns, as well as peripheral centres) but where there is now a need for their work.

An important characteristic of contemporary international migration (especially economic) is its feminization. However, this issue needs to be considered not only in terms of the growing number of women participating in migration, but also the increased interest in the subject among researchers. During the period 1960–2015, the number of migrant women doubled, but there were also increases in both the number of migrant men and the world population. The relative participation of women in 2015 compared to 1960 increased from 46.6 per cent to 48 per cent. Researchers into global migration patterns have demonstrated that women have always participated in migration, often in considerable numbers, but that they were not fully taken account of in past research. The current feminization of migration also relates to an increase in awareness of the characteristics of female migration and investigation of the causes and consequences of migration gender balance, which varies over time (Donato and Gabaccia 2015).

The increase in the number of women participating in migration has been particularly evident in Asia, where before the end of the 1970s few women migrated to find work. Since that time, there has been a surge in demand for female domestic workers, first in the Middle East, and since the 1990s, within the rest of Asia.

An interesting form of female migration in the Asian countries is marriage migration. It should be emphasized that migration for the purpose of marriage is one of the few forms of permanent migration available in the region (most Asian migration is temporary). Asian women migrated as brides of American soldiers from the 1940s onwards, first from Japan, then Korea and Vietnam. In the 1980s, there was the phenomenon of Asian 'mail-order brides' in Europe and Australia. Since the 1990s, farmers in rural Japan and Taiwan have been finding marriage partners abroad due to the exodus of local women to the cities in search of more attractive living conditions. Early in the twenty-first century, brides for men in India were recruited in Bangladesh, and Chinese farmers sought wives in Vietnam, Laos and Burma. Today, the migration of women from developing countries in Asia such as China, the Philippines and Vietnam for the purpose of marriage to men from richer Asian countries such as Japan, South Korea, Singapore and Taiwan is the most dynamic form of permanent migration in East Asia (Yang and Chia-Wen Lu 2010).

People of working age have always predominated among international migrants, whose average age in 2015 was thirty-nine years old, up slightly from thirty-eight in 2000. It is worth noting that we can see regional differences in this regard: the average age of migrants in Europe and North America is increasing, whereas in Asia, Latin America and the Caribbean, it remains low.

The number of migrants who are unaccompanied minors has grown on an unprecedented scale in recent years. Of those seeking asylum in the European

Union in 2015, 88,300 were unaccompanied minors, while in 2008–2013, this number ranged between 11,000 and 13,000 annually. In the United States from the beginning of 2014 to the end of August 2015, U.S. Customs and Border Protection apprehended more than 102,000 unaccompanied children at the border with Mexico (up to 2012 this number did not exceed 20,000 annually). Among those apprehended, 76,000 came from Central America (mainly El Salvador, Guatemala and Honduras) and 25,000 from Mexico. Children constitute a uniquely vulnerable migrant population. They travel alone to join family members who are abroad in order to escape persecution in their home country, or as victims of human trafficking, smuggling or gang violence. Although there is consensus that children travelling alone should be treated in a manner appropriate to their age, ensured protection and offered residence opportunities, the fact that young, 15–17-year-old men constitute a significant proportion of the migrant population has caused quite significant controversy. There have, for example, been voices calling for these young people to be treated in the same way as adults and a more restrictive approach taken to their applications for protection. In light of modern trends in migration, we can assume that there will be continuing growth in the number of unaccompanied migrant children, and how to help and integrate them will be a major challenge for host countries.

In the context of the diversity of migration and the varying motives behind decisions to migrate, we should also briefly discuss a process occurring primarily in highly developed countries: the growing scale of what is called environmental preference migration (Williams, King and Warnes 1997). This relates to individuals for whom the main motive of migration is not economic, but rather the quality of life and aesthetic considerations. It should be noted that here the decision to migrate is taken by people who are wealthy and independent, and often retired. This type of lifestyle migration involves moving to a place set in agreeable rural surroundings or with a sunnier climate where a more pleasant and healthy lifestyle can be enjoyed. There are a number of variants of such migration. There are those who want to 'escape to the sun' and live in resorts on the Mediterranean coast of Spain, others are 'international counterurbanisers' such as people from the UK who buy homes in the French countryside, and 'countercultural migrants' like the Germans and Danes, who settle in Ireland to lead 'alternative' rural lifestyles (King 2002). The example of European migrants who spend their time in the sunny south well illustrates the fact that the boundary between migration and other forms of mobility is difficult to capture. The range of people movements includes tourist trips (see Box 1.3) along with seasonal and permanent relocations to attractive areas. The scale of lifestyle migration is difficult to determine on a global scale as we do not have data concerning the motives for migration, but it is, nevertheless, an interesting phenomenon which highlights the multidimensional nature of the migration process.

The fourth characteristic of contemporary migration is the *shift in migration destinations*, which is connected with *the proliferation of migration transition*. The concept of migration transition derives from the hypothesis proposed by Wilbur Zelinsky (1971) that changes in social mobility occur in parallel with the

Box 1.3 Tourism

When looking at modern migration, the focus should not be exclusively on the subjects most frequently discussed in political debate, namely, economic migration and refugee issues. No less important in today's world is tourism, a form of mobility which in recent years has undergone significant changes.

Tourism is one of the fastest developing areas of the modern global economy. Its rate of growth is not far behind that of the IT industry, whose services it liberally uses. This huge dynamism, expressed in terms of the number of tourists and volume of turnover, is linked with rapid structural changes. Organizational forms, ownership structures, distances travelled, modes of travel and destinations, as well as tourists' home and host countries are all changing.

For various reasons, including the imprecise definition of the concepts 'tourism' and 'tourist', the imperfection of the statistics and inconsistency of data gathering methods relating to tourism, it is difficult to accurately determine the dimensions of tourism. However, many scholars agree with estimates that little more than a century ago, on the threshold of the First World War, the number of people travelling for tourist purposes did not exceed 20 million, and that the slow growth in tourism during the interwar period was halted by the Second World War (in 1950, some 30 million). In the 1950s and subsequent decades, however, this increase accelerated (1960: 100 million; 1980: 300 million; 1990: 415 million; 2000: nearly 700 million). Significant factors in this change were the growing prosperity of the populations of the highly developed countries and the development of modern means of transport (the first long-haul flight was made in 1969 by a Boeing 747). Even more impressive is the dynamism of change since the year 2000. In the following fifteen years, the number of tourists almost doubled, and this despite potentially adverse phenomena, terrorism in particular. Among the causes, but also the effects, of the boom in tourism, we cannot overlook the ongoing process of globalization in all its various aspects.

The mass character of tourism and the associated prospect of significant profit have led to the expansion of the tourism industry, with the largest players – major travel agencies (growing as a result of mergers and acquisitions), hotel chains and airlines (including low-cost) competing on quality and price, but the supply of destinations has also expanded. An increasing number of territories have begun to see tourism as a means of attracting wealth, and tourists, often bored with their experience of 'traditional' areas, have begun to venture to increasingly distant parts of the globe – from the South Pacific islands to the Far North, the almost uninhabited Patagonia or the 'unspoiled' Antarctic. In addition, the hitherto predominant forms of tourism focused on relaxation, including the three (or more) S's ('sea', 'sun', 'sand'), and discovery, have been joined by a wide variety of more specialized forms of tourism, and also health tourism (as well as, more controversially, travel for the purposes of euthanasia and abortion).

The development of mass tourism, often 'enclave' in character, in which tourists are, in fact, relatively isolated from the real world in the host areas, has been accompanied by an increase in the number of individual travellers for whom decreasing travel costs means the easier realization of their dream to discover the world. The growth in the number of the world's tourists has been assisted by the emergence of newcomers onto the 'tourist supply market', especially the populous and prosperous countries of East Asia. However, when assessing the dynamic contemporary development of tourism, we must not overlook the accompanying adverse phenomena – the increased threat to the natural environment along with negative social and cultural changes in host areas.

phases of demographic and economic change (the development from traditional to highly developed societies). Demographic and economic development is accompanied by changes in the destinations and intensity of migration flows (King 2012; Skeldon 2012). Currently, more countries are experiencing not only internal migration but also international migration, and their significance as both emigrant and immigrant countries is rapidly growing.

Europe, formerly the main source of emigration, has become the most important reception area. The percentage of migrants of European origin in countries which traditionally receive migration such as the United States, Canada, Australia and New Zealand is falling, while there is an ongoing rise in the number of migrants arriving in those countries from Asia in particular. In 2015, of the 244 million international migrants, 104 million came from Asia (43 per cent), 62 million from Europe (25 per cent), 37 million from Latin America and the Caribbean (15 per cent) and 34 million from Africa (14 per cent). In 2015, according to UNDESA, the countries with the largest number of persons who were living outside their major area of birth included: India (16 million), followed by Mexico (12 million), Russia (11 million), China (10 million) and Bangladesh (7 million). The proliferation of migration transition is taking place today when traditional countries of origin become hosts. The increase in transit migration (i.e. an influx of people into one country in order to reach another) is often the first step to a country being transformed from a country characterized by emigration to one which receives immigration (as is happening in e.g. Spain and Turkey) (Castles, de Haas and Miller 2014). Figure 1.2 shows the traditional areas of immigration and emigration, as well as new regions where there has been an increase in the number of international migrants. These may potentially become reception regions.

The fifth feature of migration today is its *politicization*. The context is that the issue of migration is a key subject of interest to politicians and decision makers. The creation of policy proposals to address the challenges associated with migration has inevitably become an element in political rivalry and conflict. Politicization is the result of the conviction on the part of politicians and the public that migration has critically important consequences for the security and development of the international community. On the one hand, we can see that in recent decades migration is no longer possible to control only by the internal legislation of individual countries and it has become subject to numerous regulatory measures on a transnational scale, including by the United Nations and European Union. On the other hand, national policies, bilateral agreements and regional relations are increasingly being modified under the influence of migration (e.g. Turkey–EU relations).

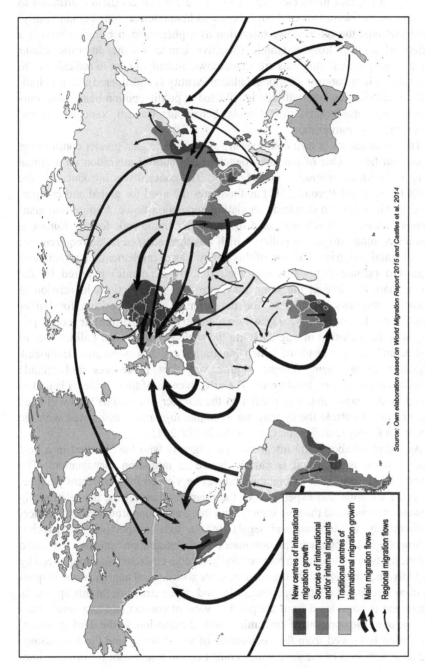

Figure 1.2 Traditional and new target regions of international migration.

Source: Own elaboration based on *World Migration Report 2015 and Castles et al. 2014*

Legend:
- New centres of international migration growth
- Sources of international and/or internal migrants
- Traditional centres of international migration growth
- Main migration flows
- Regional migration flows

The politicization of migration has been spurred by the growing number of refugees in migrant flows (see Box 1.4). There are two conflicting attitudes to this specific problem. One of them is based on humanitarian 'one human family' considerations, the other views migration as a phenomenon bringing with it a variety of threats, from economic (negative impact on the domestic labour market) and political (migrants are an unknown quantity, prone to radicalism), to cultural (the host country's national cultural identity is endangered). Within individual countries, the latter attitude is reflected in public opinion being unfavourable towards immigrants, radical social behaviours (racism, xenophobia) and laws restricting immigration.

The governments of host countries are attempting to gain greater control over population flows. One of the ideas developed to control migration on a global scale is known as 'migration management'. The concept was introduced in the 1990s (Geiger and Pécoud 2010) and assumes the need for global and comprehensive principles and standards regulating population flows. A migration management based approach was used to create the framework for the European Union's common migration policy, which was then adapted to the requirements of individual countries. The aim of this policy was to implement the principle of 'regulated openness' towards migration (mainly economic) required by the labour market, while at the same time introducing restrictions in relation to unwanted migrants. Parallel to the development of mechanisms for border control (in relation to entry, stay within a country and departure), a second priority was the inclusion of migrants into the socio-economic and political life of host countries, as a result of which integration policies were also developed. Migration management systems require very fast procedures and reliable information concerning local needs. The effectiveness of these systems is undermined by the impossibility of predicting the number of migrants who will need to be received outside the system, for example, migrations associated with the process of family reunification (Jaczewska 2013a, 2015).

As indicated above, the integration of migrants who have settled in a host country is an issue which is closely linked to migration management. The concept of integration is ambiguous and can be viewed from different perspectives: the situation and experience of the migrants themselves; the relationship between migrants and the host society; and integration opportunities and barriers (both mainly institutional and legal). In the first perspective, the psycho-anthropological approach predominates and integration means the most 'balanced' type of adaptation to a culturally diverse environment experienced by the individual immigrant. In the second, more sociological approach, the emphasis is on intergroup relations, social order and social structure. In this approach, integration is considered from the point of view of conflict and harmony within society, or as the opposite of discrimination and exclusion. In the third approach, integration is viewed from the perspective of social policy and the institutional and legal activities of the state which constitute the framework for migrant integration and the instrument for shaping intergroup relations. The latter two of these perspectives are most often reflected in media reports (Jaczewska 2013b,

Box 1.4 The European migration crisis

The dramatic increase in forced migration in the last dozen or so years, prompted by conflicts and also disasters, has made the world aware that we are dealing with a *global refugee crisis*. According to data from the UN Refugee Agency (UNHCR), the number of *forcibly displaced people* in the world reached a record 65.3 million in 2015 (59.5 million in 2014 and 43.7 million in 2001). Of these, 21.3 million in 2015 were refugees (19.5 million in 2015, 15.4 million in 2010) and 10 million were *stateless*. Among refugees worldwide, Syrians are today the most numerous group (4.9 million in 2015, and 3.9 million in 2014), followed by Afghans (2.7 million in 2015), who had constituted the largest global refugee population in the three decades up to 2014, and Somalis (1.1 million). In 2015, these three countries 'supplied' a total of 54 per cent of the world's refugees. It is worth noting that it is developing countries (often those adjacent to crisis areas) which host the largest share of refugees (86 per cent in 2015), and this includes the world's least developed countries, where 26 per cent of refugees find shelter in 2015. Turkey, at the beginning of 2016, hosted the largest number of refugees in the world (2.3 million, compared to 1.6 million a year earlier), followed by Pakistan (1.6 million), Lebanon (1.1 million) and Iran (1 million).

The European migration crisis (or European refugee crisis) is the name given to the rapid increase in the number of migrants coming to the EU via the Mediterranean and south-east Europe in search of a better life. Although the number of migrants has been increasing for several years, the beginning of the crisis is considered to be the year 2015, when a record 1.3 million asylum applications were made in the EU (double that of the previous year).

The European migration crisis should be seen as part of the global refugee crisis. An extremely important factor in this from a European perspective is the wave of social unrest, which began in Tunisia in December 2010 and spread to the Arab World (the 'Arab Spring'). Relatively limited migratory effects have resulted from the tensions in Tunisia and Egypt, but the conflicts in Libya and Syria have generated a large stream of refugees (among them African and Asian migrant workers employed in the region).

The unstable situation described above, along with ineffectual EU migration policies, have led to large-scale flows of African, Asian and also European migrants seeking refuge and a higher quality of life in the EU countries. In 2015, more than 1 million people tried to cross the Mediterranean in order to enter the EU, of whom 848,000 and 153,000 reached Greece and Italy respectively (more than 3,700 died making the attempt).

Migrants reach Europe by a variety of routes, which change when countries seal their borders (see Figure 1.3.). In 2015, the most commonly used was the western Balkan route running through Turkey, the Aegean Sea, Greece (mainly the islands of Kos, Chios, Lesbos and Samos), Macedonia and Serbia to Hungary. There are also other routes (based on data from Frontex[2]), which have been used for many years: the West African, Western Mediterranean and Central Mediterranean routes (the last of these was the route most often used in 2014), also a route through Apulia and Calabria, and one from the eastern border of the EU. In September 2015, a new Arctic route running from Russia to Norway and Finland came into use.

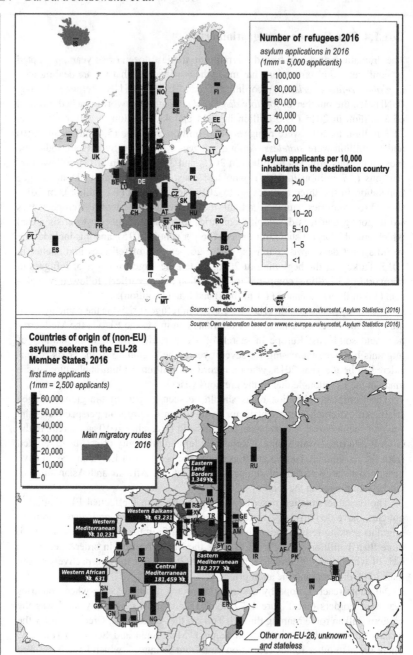

Figure 1.3 The number of refugees and countries of origin of (non-EU) asylum
seekers in the EU-28 Member States in 2016.

The European migration crisis has made politicians aware not only of the scale of potential migration into Europe and the determination of migrants to enter Europe, but also of the lack of unanimity and solidarity within the European Union, and its ineffectual handling of the crisis. It has not yet been resolved and the European Commission has been making what are thus far insufficient attempts to stabilize the situation (refugee relocation programmes within the EU, increased controls on the EU's external borders, negotiations with third countries including Turkey aimed at preventing further waves of illegal migration, measures to counteract organized crime and people smuggling, attempts to stabilize areas where large numbers of refugees have already been received). The European Commission estimates that from the beginning of 2016 to the end of 2017 as many as another 3 million migrants and refugees may arrive in the EU.

2015). Integration policy has become the 'political slogan of the day', but it also remains a significant challenge for most countries which have accepted, are accepting and intend to accept migrants.

Documents of international institutions (such as the International Organization for Migration and the European Commission), define integration as a two-way process by which immigrants become an accepted part of society. Integration assumes that migrants and hosts will respect all existing laws and obligations as well as values linking migrants and the host society. Although there is agreement as to the validity of understanding integration as a two-way process, there are somewhat large differences as to the depth of the adjustment to be made by both sides and the effect to be achieved (e.g. whether the principle of multiculturalism should prevail or not). That approaches to migration policy are diverse stems from disparate political and ideological perspectives, the variety of policy mechanisms and instruments, the history of immigration in a given country and the social situation of the immigrants themselves. A characteristic feature is the fact that in contemporary societies, their attitudes towards immigrants and their philosophies of integration are constantly evolving under the influence of changing social realities, and political and economic priorities.

The problems which are currently being experienced by host countries stem from the failure of a large number of the policies implemented thus far. One of the many examples of this is the creation of 'parallel societies' where the representatives of different nationalities live next to each other and not together. Governments are faced with the challenge of how to ensure social cohesion in a context of extreme social, ethnic and cultural diversity. Numerous tensions between 'traditional' communities in the host countries and first, second and even third generation immigrants have found their outlet in a growing number of public disturbances (e.g. in the suburbs of Paris), assaults (especially on Muslims) and even terrorist acts.

A new field of research into migration relates to the mass media.[3] On the one hand, these media have begun to influence migration decisions, since they allow migrants to pay virtual visits to potential destinations, and more easily make arrangements for travel, accommodation and work, and then when in a host country to maintain contact with their own community and transnational networks. On the other hand, they have an influence on how migrants are perceived. This role of the mass media should not be underestimated, for in the quest for sensational news, the media can create a negative image of migration, or, by contrast, myths can be dispelled and integration policies promoted through information campaigns in the media countering xenophobic behaviour and racism etc.

A distinguishing feature of our times is the global nature of migration, whose patterns are rooted in history and whose contemporary contours are affected by various political, demographic, socio-economic, geographical and cultural factors. The consequence of these migration flows is a greater ethnic diversity within countries and deepening transnational links between countries and communities. The policies of individual governments and international bodies have a huge impact on migration. We can identify several contemporary dilemmas whose outcomes will be decisive for the future shape of our societies, as well as relations between countries. These include the following stark alternatives, should we: encourage the further development of global migration, or instead seek to curb it?; improve cooperation and governance structures in relation to migration, or abandon existing policy objectives?; adopt measures to stem illegal migration, or simply ignore it?; regulate legal migration and integrate people who want to settle, or 'close the door' to all newcomers?; and, finally, should we embrace, or not, the role of ethnic diversity in social and cultural change and the consequent transformation of our modern nation states?

Conclusions

Several thousand years ago, in the Book of Genesis, the commencement of the Jewish Pentateuch, an unknown author recorded these words of Yahweh addressed to human beings: 'Be fruitful and multiply, and fill the Earth and subdue it'. Looking at human history from the perspective of the twenty-first century, we can only conclude that this precept has already been completely fulfilled. The number of people on Earth is without precedent in the history of the species. There is no place in the world where you cannot meet representatives of *homo sapiens*. We are increasingly transforming our planet. Therefore, there can be no questioning the legitimacy of current attempts to devise a name for our times which is both commensurate with humankind's present role in the Earth's ecosystem and in some way reflects it linguistically. Hence, the appearance of such proposals as Anthropocene (from Greek *anthropos* 'human being' + *kainos* 'new') (Castree 2015).

Notes

1 Demographic issues are strongly associated with the use and interpretation of numerical data. Given the limitations which arise from the ways in which population is recorded, and the number of variables which inevitably affect their credibility, we believe that it is justified to ask whether or not it is possible, and indeed worthwhile, to describe demographic reality without the use of numerical data.

In a 2008 article titled *Truth, Damn Truth, and Statistics*, Paul F. Velleman, professor of statistics at Cornell University, addresses the importance of ethics in statistics used in academic research and everyday life. The title of the article is a clear reference to the quotation attributed by Mark Twain to the British politician and writer Benjamin Disraeli (1804–1881): 'There are three kinds of lies: lies, damned lies and statistics'. This is surely a warning to those who have boundless trust in numerical data on the basis of which they construct categorical and, to their minds, unambiguous truths.

Theoretically, therefore, demographic issues can be presented without statistics, but this would be just as much a manipulation as when statistics are used naively. Thus, any analysis of demographic trends in the modern world should maintain a certain harmony between the use of statistics and an awareness of their imperfection, which can be commented on in the descriptive layer.

2 Frontex (*Frontières extérieures*, French for 'external borders') is an agency of the European Union with its headquarters in Warsaw, Poland, working with the border and coast guards of Schengen Area member states.

3 The term 'mass media' refers to a diverse range of media technologies which reach a large audience by means of mass communication. This takes place through a variety of channels, for example, broadcast media, which transmit information electronically via film, radio, recorded music and television, and digital media which comprise both internet and mobile mass communication.

References

Asylum Statistics (2016). Retrieved 15 April 2017 from http://ec.europa.eu/eurostat.

Bauman, Z. (2000) *Liquid Modernity*. Polity: Cambridge.

Black, R., Engbersen, G., Okólski, M. and Panţîru, C. (2010) *Continent Moving West? EU Enlargement and Labour Migration from Central and Eastern Europe*. IMISCOE Research, Amsterdam: Amsterdam University Press.

Caldwell, J.C. (2006) *Demographic Transition Theory*. Dordrecht: Springer.

Carey, H.C. (1867) *Principles of Social Science*, Philadelphia, PA: J.B. Lippincott & Co; London: Trubner & Co; Paris: Guillaumin & Co.

Castles, S., de Haas, H., Miller, M.J. (2015) *The Age of Migration, International Population Movements in the Modern World*. London: Palgrave Macmillan.

Castree, N. (2015) The Anthropocene: A Primer for Geographers. *Geography* 100(2): 66–75.

Donato, K.M. and Gabaccia, D. (eds) (2015) *Gender and International Migration: From the Slavery Era to the Global Age*. New York: Russell Sage Foundation.

Geiger, M. and Pécoud, A. (eds) (2010) *The Politics of International Migration Management*. London: Palgrave Macmillan.

Jaczewska, B. (2011) A World City or the World in One City. A Case Study of London, in M. Czerny and J. Tapia Quevedo (eds), *Metropolitan Areas in Transition*. Warsaw: Wydawnictwo Uniwersytetu Warszawskiego.

Jaczewska, B. (2013a) Migration Management in the European Union: Immigration Policy in Germany and the United Kingdom, in M. Banaś and M. Dzięglewski (eds), *Narratives of Ethnic Identity, Migration and Politics: A Multidisciplinary Perspective*. Kraków: Księgarnia Akademicka.

Jaczewska, B. (2013b) Multidimensionality of Immigrant Integration Policy at the Local Level: Examples of Initiatives in Germany and the United Kingdom. *Miscellanea Geographica* 1(17): 25–33.

Jaczewska, B. (2015) *Zarządzanie migracją w Niemczech i Wielkiej Brytanii: Polityka integracyjna na poziomie ponadnarodowym, narodowym i lokalnym*. Warsaw: University of Warsaw, Wydział Geografii i Studiów Regionalnych.

Kalaska, M. and Wites, T. (2015) Perception of the Relations Between Former Colonial Powers and Developing Countries. *Third World Quarterly* 36(10): 1809–1826.

King, R. (2002) Towards a New Map of European Migration. *International Journal of Population Geography* 8: 89–106.

King, R. (2012) Geography and Migration Studies: Retrospect and Prospect. *Population, Space and Place* 18: 134–153.

Konarzewski, M. (2005) *Na początku był głód*. Warsaw: Państwowy Instytut Wydawniczy.

Kremer, M. (1993) Population Growth and Technological Change: One Million B.C. to 1990. *The Quarterly Journal of Economics* 108(3): 681–716.

Malthus, T.R. (1798) *An Essay on the Principle of Population As It Affects the Future Improvement of Society, with Remarks on the Speculations of Mr. Goodwin, M. Condorcet and Other Writers*, 1st edn. London: J. Johnson in St Paul's Church-yard.

Meadows, D.H., Meadows, D.I., Randers, J. and Behrens, III W.W. (1972) *The Limits to Growth: A Report to the Club of Rome*. Washington, DC: Potomac Associates.

Notestein, F.W. (1945) Population: The Long View, in T.W. Schulz (ed.), *Food for the World*. Chicago, IL: University of Chicago Press, 36–57.

Okólski, M. (2010) Spatial Mobility from the Perspective of the Incomplete Migration Concept. *Central and Eastern European Migration Review* 1(1): 11–35.

Pawłowski, S. (1938) Na zamknięcie Międzynarodowego Kongresu Geograficznego w Polsce, in S. Pawłowski (ed.), *Geografia jako nauka i przedmiot nauczania*. Lwów: Książnica-Atlas.

Penning de Vries, F.W.T., Van Keulen, H., Rabbinage, R. and Luyten, J.C. (1995) Biophysical Limits to Global Food Production. *Brief*. Washington, DC: International Food Policy Research Institute.

Pickford, M. and Senut, B. (2001) 'Millennium Ancestor': A 6-Million-Year-Old Bipedal Hominid from Kenya. *South African Journal of Science* 97(1–2): 22.

Secomski, K. (1978) *Polityka społeczno-ekonomiczna: Zarys teorii*. Warsaw: PWE.

Sen, A. (2000) *Conclusions – IIRPD*. New York: Anchor Books.

Simon, J. (1996) *The Ultimate Resource: Revised Edition*. Princeton, NJ: Princeton University Press.

Skeldon, R. (2012) Migration Transitions Revised: Their Continued Relevance for the Development of Migration Theory. *Population, Space and Place* 18(2): 154–166.

Solarz, M.W. (2014) *The Language of Global Development: A Misleading Geography*. Abingdon: Routledge.

Solarz, M.W. and Wojtaszczyk, M. (2015) Population Pressures and the North–South Divide Between the First Century and 2100. *Third World Quarterly* 36(4): 802–816.

UNDESA (2016) United Nation Department of Economic and Social Affairs, Population Division. Retrieved 10 September 2016 from www.un.org/en/development/desa/population/migration/.

UNDP (2015) Human Development Report: Work for Human Development. New York: United Nations Development Programme. Retrieved 10 September 2016 from http://hdr.undp.org/sites/default/files/2015_human_development_report.pdf (accessed).

UNHCR (2016) Global Trends, Forced displacement 2015. Retrieved 10 September 2016 from www.unhcr.org/statistics/unhcrstats/576408cd7/unhcr-global-trends-2015.html.

United Nations (2005) *Population Challenges and Development Goals*. Department of Economic and Social Affairs, Population Division. New York: United Nations.

UNPF (2016) World Population Day. United Nations Population Fund. Retrieved 10 September 2016 from www.unfpa.org/events/world-population-day.

Van de Kaa, D.J. (2004) Is the Second Demographic Transition a Useful Research Concept? Questions and Answers. *Vienna Yearbook of Population Research* 2: 4–10.

Velleman, P.F. (2008) Truth, Damn Truth, and Statistics. *Journal of Statistics Education* 16(2).

Vertovec, S. (2007) Super-Diversity and its Implications. *Ethnic and Racial Studies* 30(6): 1024–1054.

Williams, A.M., King, S. and Warnes, A.M. (1997) A Place in the Sun: International Retirement Migration from Northern to Southern Europe. *European Urban and Regional Studies* 4: 115–134.

Wites, T. (2006) Frontier Delimitation of the North in Europe. *Acta Geographica Universitatis Comenianae* 49: 71–80.

Wites, T. (2007) *Wyludnianie Syberii i rosyjskiego Dalekiego Wschodu.* Warsaw: Wydawnictwa Uniwersytetu Warszawskiego.

Worldometers (n.d.a) Retrieved 10 September 2016 from www.worldometers.info/world-population/Niger.

Worldometers (n.d.b) Retrieved 10 September 2016 from www.worldometers.info/world-population/Ethiopia.

Worldometers (n.d.c) Retrieved 10 September 2016 from www.worldometers.info/world-population/Russia.

Worldometers (2017) Retrieved 10 September 2016 from www.worldometers.info/world-population/.

Yang, W.S. and Chia-Wen Lu, M. (eds) (2010) *Asian Cross-border Marriage Migration: Demographic Pattern and Social Issues.* Amsterdam: Amsterdam University Press.

Zakharov, S. (2008) Russian Federation: From the First To Second Demographic Transition. *Demographic Research* 19(24): 907–972.

Zelinsky, W. (1971) The Hypothesis of the Mobility Transition. *Geographical Review* 61: 219–249.

2 Geographies of development in the twenty-first century

Ferenc Gyuris

Preface

The twentieth century is often thought of as a period of unprecedented improvement in many spheres of life. New technologies revolutionizing communication, traffic and medicine, successful attempts to involve more people in education than ever before, the discovery of cures for dangerous diseases, the invention of production methods resulting in a variety and quality of products and services previously unimaginable, are important and very visible parts of this process. These boom-like trends in the twentieth century were, especially from the middle of the century onwards, commonly referred to as 'development'. In spite of many controversies emerging around this concept in the last few decades, post-millennium global steps to promote the quality of living of human beings all around the world are still often taken under the aegis of development, as illustrated by the Millennium Development Goals of the United Nations, for example. Furthermore, concepts challenging either the underlying notion of 'development' in the sense it was widely used in the Cold War period, or the way its implementation was attempted in practice, are often aimed at goals closely related to those which many former promoters of 'development' also wanted to achieve in the long run, such as social equality or the dissolution of poverty. Hence, this chapter focuses on the contested concept of development, its historical roots, the manifold ways it has been theoretically conceptualized and practically utilized, its inequities and the sometimes rather exclusive instead of inclusive outcomes of developmental policies. Finally, early twenty-first-century trends are discussed, as well as some opportunities for dealing with these issues from new and different perspectives.

Development: a contested concept with contested policies

The notion of development, in the sense in which it made a worldwide career in the twentieth century, was rooted in certain fundamental concepts of the European Enlightenment. Human society was widely believed to have gone through a development process throughout history, resulting in an increasing number of people 'emerging' from a state of ignorance and immorality, and attaining high

cultural standards, with 'cultural' referring to a number of aspects, from education to physical hygiene to political self-consciousness. In other words, the
history of humankind was perceived as a history of progress, the advance of
'civilization' and decline of 'barbarism'. This narrative pervaded colonial politics in the nineteenth and early twentieth centuries, where the political elites of
the colonial powers commonly justified their imperialist and racist expansionary
project as a 'civilizing mission' or *mission civilisatrice*, with white Western
peoples, claimed to be 'superior', bringing the torch of the Enlightenment to
peoples in other parts of the globe, who were considered 'savage' and 'inferior',
and whose communities were allegedly based on 'childish' principles of social
organization (Bullard 2000; Butlin 2009; Conklin 1997).

Meanwhile, the elites of the emerging European nation states widely conceived the main *raison d'état*, the reason for the existence of the state, to be the
struggle against the remnants of social problems inherited from former periods
of history and the promotion of civilization and cultural development inside the
national borders. In fact, this often ended in attempts to establish a homogenized
national culture, often at the cost of disqualifying cultures, traditions and types
of knowledge associated with ethnic, linguistic, religious etc. minorities (Meusburger 2016). Political leaders, statistical offices and even academics paid ever
more attention to the social and spatial inequalities of literacy, school enrolment
and crime (Meusburger 1998; Gyuris 2014a), and even international rivalry
between various countries was increasingly addressed as a conflict between a
'more civilized' Self and a 'less civilized' or even 'barbarian' Other. A most
typical example of this was the First World War, where countries on both sides
fought in the name of such claims, and even the narratives of the new political
order after the peace treaties reflected these concepts, both in politics and
academia (*Treaty of Peace with Germany* 1919; Schroeder-Gudehus 2014).

During this period, however, 'development' was not a commonly used term
in political and academic vocabularies with regard to differences between
various countries claimed to be at 'higher' or 'lower cultural standards' and
levels of 'civilization' (Schlüter 1908; Jefferson 1911; Huntington 1915, 1927).
The heyday of developmentalist language arrived only after the Second World
War, when the global political framework underwent remarkable changes and
witnessed the emergence of the Cold War between two superpowers, the United
States and the Soviet Union, both lying outside the European political core. In
this new context, providing economic assistance became a major strategy of US
governments to stabilize anti-Communist political regimes in existing allied
countries and especially to make new allies in parts of the world where American influence had been weak or even non-existent before, but where the emergence of Communist regimes trying to orientate their countries towards the
USSR seemed a real possibility (Gyuris 2014a; Solarz 2014). An eminent
example of this was the 'race' between the USA and the Soviet Union, especially in the 1960s, for sway over the numerous newly liberated countries in
Africa, which had previously been under the domination of various European
colonial powers.

The underlying concept of development was made explicit as early as Harry Truman's 1949 inaugural speech, where he referred to (economic) development as a major prerequisite for global peace, freedom and prosperity. He put it in these words: 'Greater production is the key to prosperity and peace. And the key to greater production is a wider and more vigorous application of modern scientific and technical knowledge' (quoted in Woolley and Peters n.d.). From then on, the United States mobilized considerable resources to give both economic aid and long-term loans to 'undeveloped' countries, especially those in geopolitically sensitive regions (Pounds 1963). In fact, support was often provided not by the US government directly, but through US dominated international institutions such as the World Bank, which was established in 1944 as the International Bank for Reconstruction and Development (IBRD) and headquartered in Washington, DC. These funds were mostly channelled to development projects aimed at laying the foundations of modern industrial activities and creating related infrastructure. Such projects in general were modelled on the TVA (Tennessee Valley Administration) initiative launched in the United States in 1933 in order to enable economic restructuring, stimulate growth and create new workplaces in a region that had been suffering from a deep crisis in traditional mining in the nearby Appalachian district (Ekbladh 2002). In the same way as the TVA, which Scott (1998: 6) refers to as the 'granddaddy' of development projects throughout the world, the construction of huge hydropower plants and the improvement of energy production in order to serve the needs of modern heavy industry were the cornerstones of development projects in many countries gaining firm US support for geopolitical interests (Ekbladh 2010; Sneddon 2015).

In fact, this technological and economic interpretation of development was in line with US geopolitical concepts, which rejected the traditional European practice of colonization and conceived of the expansion of the American sphere of influence as taking place primarily through economic projects (Agnew 1993; Dorband 2010). This sort of involvement was also more acceptable to the former colonies, which now as sovereign states still had the burden of a colonial heritage and usually rejected projects reminiscent of European 'civilizing missions'.

A massively technology and economy-oriented interpretation of development was also compatible with, and thus further justified by, official views in the Soviet Union. Marxist-Leninist (and, in the early Cold War period, Stalinist) concepts also interpreted human history as a process of linear progress with societies gradually rising from 'inferior' to 'superior' modes of production, with socialism and then communism claimed to be at the top. The economy was believed to be the basis for the social 'superstructure', meaning that social development was dependent on the development of economic production (Niedermaier 2009). Furthermore, 'actually existing socialist' regimes also saw themselves as contexts where, unlike in capitalism, rapid modernization did not face obstacles resulting from the (counter-)interests of individual companies and investors (Stalin 1972 [1952]; Ostrovityanov *et al.* 1957 [1954]). The Soviet regime, therefore, continuously launched and carried out large-scale development projects within its own boundaries, especially in remote regions of Soviet

Asia and what is known as the 'Far North' (Gyuris 2014a; Klüter 2000), in its satellite countries, and in a number of 'friendly socialist' countries in Africa and Asia, and even in Latin America (Cuba) (Brun and Hersh 1990; Hong 2015). Although becoming part of the Soviet sphere of influence meant for countries in East Central Europe that a massive amount of resources were sucked out by the USSR (Bunce 1985; Marer 1974, 1976), newly liberated countries in the 'Third World' often gained real and significant economic support from the Soviet Union. Due to the weakness of the Soviet economy compared to its US counterpart, the total global volume of such economic assistance was not on a par with that of America. Yet, for many poor countries it still meant considerable support, especially if it was, at least in a geopolitical sense, carefully focused on a few especially sensitive areas (Pounds 1963).

Like those of the Western Bloc, Soviet-style development projects also concentrated on economic development in terms of promoting production. In fact, both US- and Soviet-led projects emphasized an array of issues, including improving health conditions or education levels, but regarded industrialization and economic modernization as the main engines of the process of development. That is why the measurement of development, which initially lacked feasible indicators as well as the statistical framework to produce them, gradually moved towards quantifying economic production. This opened the door to the introduction of the Gross Domestic Product (GDP) indicator, which had a rapid international rise in the Western Bloc as the *par excellence* indicator of development (Dickinson 2011). In the Soviet Union's economic statistics, instead of GDP, the Net Material Product (NMP) indicator was used, which, since Marxist-Leninist principles defined production as a *material* process, did not include non-material or, in the language of the day, 'non-productive' activities and their outcomes, which mostly belonged to the service sector (Campbell 1985). Yet, both indicators of development had a firm focus on economic production and were compatible with the notions of development then prevailing. Widely used proxy indicators such as the share of industrial employment, or per capita industrial production also reflected the great importance that economic conditions were believed to have in relation to development in general.

From the late 1970s onwards, however, mainstream thinking about development gradually changed. Until then, development projects had been motivated by a Fordist understanding of the economy, with an emphasis put on heavy industries, economies of scale, mass production of standardized goods and relatively strong state regulation. In the meanwhile, factors such as energy efficiency or the efficient use of raw materials gained little attention due to the low prices of these products in the international market. Environmental issues were similarly neglected. Instead, large-scale construction projects resulting in a massive transformation of nature were regarded as signs of 'progress' and increasing human mastery over nature. This was true not only for the Soviet Bloc and especially the USSR (Brain 2010; Györi and Gyuris 2015), where such initiatives were on the agenda until the late 1980s, but also in the Western Bloc. Even there, such initiatives as deforestation programmes in the Amazon rainforests, or large dam construction projects were praised for decades as developmental mile-stones representing the progress of humankind in the modern period (Andersen, Granger, Reis, Weinhold and Wunder 2002; Ekbladh 2002). Funding schemes, especially in a number of less wealthy countries in Latin America, Africa and Asia, were mainly reliant on long-term international loans taken out by national governments on favourable terms.

However, the oil crises of the 1970s as well as green movements in the Western world, which emerged in the wake of the social and political change movements of 1968, resulted in a reconceptualization of the notion of develop-ment. In parallel, the neoliberal shift in the Anglophone world exemplified by the Thatcher government in the UK and the Reagan administration in the USA pushing for deregulation and privatization, in critical literature commonly desig-nated as neoliberalization (Harvey 2007; Ward and England 2007), led to new approaches to development policies. Instead of large-scale top-down investment projects carried out according to detailed plans, structural adjustment became the new buzzword (Peet and Hartwick 2009; Rauch 2009). To foster development, international organizations and development agencies such as the World Bank increasingly urged national governments to dismantle uncompetitive activities and concentrate resources in more efficient economic sectors. Furthermore, espe-cially since the 1990s, these steps have been achieved less on the basis of favour-able international loans, but rather by decreasing alternative expenditures through austerity and raising surplus financing through privatization.

Measuring development has also undergone changes. Moving beyond a nar-rowly economic interpretation, in 1990, the United Nations adopted the Human Development Index (HDI) in order to give greater weight to factors which are more social than economic, namely education (initially measured by the adult literacy rate and gross education enrolment ratio, more recently by the mean and expected years of schooling) and health (life expectancy) (UNDP 2009, 2010). In the last roughly two and a half decades, a variety of other indicators have also emerged, opening the way for increasing attention to, for instance, environ-mental or gender equality aspects (OECD 2008; UNDP 2015).

Development for whom? The notions of spatial inequality and injustice

Since antiquity, the issue of social inequality has always been a central issue in social and political thinking. Although the degree of stratification of society has changed from period to period and place to place, and disparities have become manifest in different ways in different contexts, the concepts, ideologies and political struggles arising in response to this stratification have all seemed to take up one of two conflicting views. The first is based on the claim that all humans are equal, an idea present both in traditional Christian teachings, which interpret every human being as created by God in his own image, as well as in key con-cepts of the Enlightenment, which attribute certain inalienable rights to every single human. This notion has fuelled many diverse political agendas over history, including those aimed at universal suffrage, the redistribution of economic resources in favour of the poor and even the collectivization of property. By contrast, the second approach argues for the axiom 'to equal merit, equal reward', thus, that equal contribution to the well-being of society should result in equal honour. This implies that individual merits are different, which necessarily and 'righteously' leads to social disparity, for example, in terms of income and property (Gyuris 2014a).

In fact, real life is far too complex to organize society according to such simplistic principles. Presumably, very few would argue for total equality to such a degree that courts should let mass murderers retain the same rights as law-abiding citizens instead of sending them to jail. Meanwhile, if 'to equal merit, equal reward' is accepted as the ultimate principle, the question will arise as to what should be acknowledged as merit and who has the right to decide about the issue. Even the most cruel and inhumane totalitarian systems, such as those of Hitler and Stalin, claim to give everyone what they 'deserve'. Hence, both principles have a limited radius and a lot of blurred conceptual details. This, of course, does not invalidate, in an actual situation, for example in a debate about specific policies to implement, the legitimacy of either arguing for less inequality even at the cost of giving less reward to some, or arguing for more stimulation by rewards even if it widens the social gap. Yet, due to their enormous mobilizing power, the two principles are often taken up by various political actors as mere buzzwords. In such cases, speaking about, for example, inequality serves less as an attempt to collect various views and better understand why inequalities emerge and how they should be handled, and more as a political discourse, in which taking sides helps an actor easily express his/her views, interests and goals, also in order to find and mobilize supporters (Gyuris 2014a).

Given that an eminent form of social disparity is inequality between people in different geographical locations (different neighbourhoods, cities, regions, countries, continents and so on), spatial inequality is a very sensitive, highly political and politicized issue. As early as the nineteenth century, pioneer works on the spatiality of 'moral standards', mainly based on statistics on literacy and school attendance on the one hand, and crime on the other, revealed remarkable spatial inequalities in both educational and social deprivation between regions of the same country, as well as between various neighbourhoods of the same city (e.g. Dupin 1826; Balbi and Guerry 1829; Guerry 1833, 1864; Parent du Châtelet 1836; Booth 1902–1903). Related studies in such different countries as France, Britain, the United States and Canada soon became key reference works for various political actors wanting to implement specific policies focused on specific social groups in specific areas. Such initiatives often centred around the improvement of access to education and more rigorous legal regulation of 'immoral' behaviour, including language use in public spaces and prostitution (Hunt 2002). Furthermore, an intensively used argument to justify European colonial endeavours in Africa and Asia, or comparable US activities in the early twentieth century in Latin America especially, was that of being a 'civilizing project' (Kramer 2002), which brought 'the torch of the Enlightenment' to 'uncivilized people', thus closing the gap between lands formerly inhabited by 'superior' and 'inferior' types of human (Butlin 2009).

The classical Marxist tradition which emerged in the mid-nineteenth century explicitly pointed to the urban–rural divide (Marx and Engels 1998 [1848]) and the uneven geographical development of countries and regions (Luxemburg 2003 [1913]; Lenin 1964a, 1964b) as problems the envisaged socialist overturn was expected to solve. After the Great Depression of 1929, US and British governments also perceived the spatial inequality of development as a problem in its own right. The establishment of the Tennessee Valley Authority in the United States in 1933 was a conscious reaction to the fact that massive economic depression in a given region can be highly detrimental for the entire economy of a nation (Ekbladh 2002). Similarly, the report of the Barlow Commission in the United Kingdom in 1936, which urged the spread of industry into unindustrialized areas, was also aimed at avoiding a future in which the decline of a dominant industrial area like the Black Country might pose a threat to national development (Gilbert and Goodman 1976).

Spatial inequality of development received previously unprecedented political attention in the Cold War period. In a new global context where the European colonial empires were on the verge of collapse, the United States and the Soviet Union, as it was already presented, launched a race to gain new allies among the newly liberated countries. Each superpower made great efforts to convince the local elites that its economic and social model would bring more benefit than that of its rival. Soviet concepts emphasized the claim that global development disparities are necessary outcomes of the capitalist system, even under the domination of the United States instead of Britain. By contrast, they radiated massive optimism about the capacity of Soviet-style communist regimes to solve this problem and bring the former colonies up to a par with the more industrialized countries of the world. These views were disseminated especially energetically after the publication of Stalin's work 'Economic Problems of Socialism in the U.S.S.R' (Stalin 1972 [1952]), which was foreseen as a key economic textbook for the new local intelligentsia in the 'Third World' (Pollock 2006).

In the United States, academic works on economic development as well as public discourse did not initially consider spatial and social disparity to be a major problem, but rather a natural and growth-stimulating outcome of economic competition, the hallmark of capitalism (Glasmeier 2002). The first neoclassical concepts to pay attention to the issue did not go beyond claiming that unequal growth was a result of the lack of free movement of production factors, caused by improper state intervention (Harris 1957; Borts and Stein 1964). Since a free market economy without such 'distortions' was expected to automatically bring forth balanced growth in the long run, the adoption of the US economic model by former colonies was thought to be a simple but efficient cure for development disparities.

From the late 1950s onwards, it had become obvious to many scholars, however, that such a policy was unable to decrease global inequalities or even to keep them at a constant level (Myrdal 1957). This recognition opened the way for the emergence of polarization theories, which supposed that global and regional disparities of development could be overcome only by massive development projects coordinated by state or international organizations. These theories were also based on the idea that increasing inequality in the initial phase is an unavoidable concomitant of development, which then produces the resources for also mobilizing those regions which benefitted less from economic 'take-off' (Rostow 1960) in the early period (Myrdal 1957; Hirschman 1958; Williamson 1965; Friedmann 1966). In reality, such ideas were based more on wishful thinking and hypothesis than on empirical evidence from 'underdeveloped' or 'developing' (Solarz 2014) countries, partly because not even the most basic economic indicators in those regions had yet been measured. Yet, since these theories were in line with the political interests behind the development projects (Gyuris 2014a; Moran 2005), they contributed significantly to justifying developmentalism, and to the fact that its 'golden age' brought much less progress in closing the global and regional development gaps than initially supposed. It is important to underline that the 1950s and 1960s were the heyday of similarly designed development projects in the capitalist 'First' and the communist 'Second' worlds where they did mostly result in decreasing disparities between various regions. At the same time, however, they usually increased inequality between settlements of various sizes and functional complexity, significantly benefitting large urban centres while inducing or accelerating downward trends in small towns and villages in the rural peripheries (Meusburger 1997; Gyuris 2014a, 2014b).

From the late 1970s and 1980s onwards, regional inequality of development increasingly disappeared from the agenda of leading US headquartered international organizations, and in most countries of the world gradually lost its eminent position among national policy goals as well. This mainstream 'downgrading' of regional inequality of development as a problem, however, resulted in alternative and often radical academic and political streams, such as neo-Marxists, feminists and greens, beginning to show interest in the issue. Furthermore, these groups also took up aspects of unequal geographical development, which had previously received less consideration (Harvey 2006; Smith 2008), such as gender disparities (Massey 1994; Hanson and Pratt 1995; Dickson and Jones 2006), environmental inequalities or micro-scale forms of disparity (Harvey 1973, 1978) beyond regional and global ones (e.g. between urban neighbourhoods). Moreover, in contemporary academic discourse, unequal development is not simply a matter of distribution, where the aim is to find a sort of distribution 'optimal' for everyone, but an issue of recognition and participation (Young 1990). The main concern is not simply spatial inequality but *spatial injustice*, where 'just' forms of development require that social groups with alternative normativities and 'ways of seeing' (e.g. beyond a white, Western, male, high-income class perspective) are taken into consideration and their representatives are involved in decision-making processes (Fincher and Iveson 2012; MacLeod and McFarlane 2014; Gyuris 2017).

Development and its disparities in a globalized world: early twenty-first-century trends

On a global scale, many indicators often referred to as important dimensions of development seem to confirm that international disparity has decreased between the end of the Cold War and the mid-2010s, especially since the turn of the millennium. According to official data provided by the International Monetary Fund, the weighted standard deviation of per capita nominal GDP among countries worldwide was quite constant between 1992 and 2002, only decreasing slightly during the 1997 Asian crisis before returning to its former values. However, since 2002 we have witnessed a clear declining trend, with the standard deviation sinking from 187 per cent of the global average of per capita GDP to 142–143 per cent in 2014 and 2015 (Figure 2.1). If the unequal costs of living in different countries are considered by reference to inequalities in per capita GDP in purchasing power parity terms, the values are somewhat lower in general, for in areas with higher per capita economic productivity, prices also tend to be higher. Yet, the temporal tendency is basically the same as for

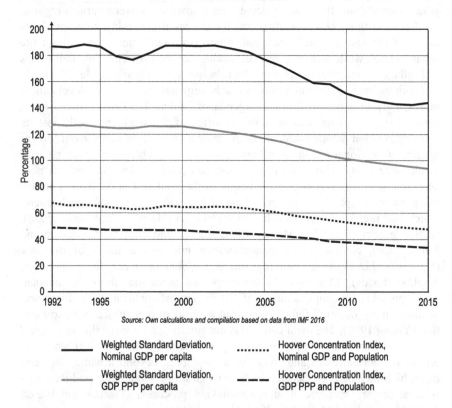

Source: Own calculations and compilation based on data from IMF 2016

| | Weighted Standard Deviation, Nominal GDP per capita | | Hoover Concentration Index, Nominal GDP and Population |
| | Weighted Standard Deviation, GDP PPP per capita | | Hoover Concentration Index, GDP PPP and Population |

Figure 2.1 Intercountry inequalities in per capita GDP, and the concentration of GDP and population in nominal number and purchasing power parity terms, 1992–2015.

nominal GDP. Disparities in the distribution of GDP and population reveal similar processes. While the Hoover Concentration Index (Huang and Leung 2009) shows that until 2002 roughly 65 per cent of total nominal GDP would have needed to be redistributed among countries to achieve international equality in per capita GDP, by 2015, this had decreased to below 48 per cent. For GDP in PPP terms, a similar change occurred, with a reduction from 46–47 per cent to 33 per cent.

For such basic health indicators as life expectancy or infant mortality, the early years of the twenty-first century, just as the latter decades of the twentieth century, seem to have brought about ongoing improvement, not only in the sense of longer lifespans and less infant mortality on average, but also in terms of a narrowing gap between more and less developed countries (Figure 2.2). According to UN official statistics, global life expectancy for both sexes has increased since the first half of the 1990s from 64.5 years to 70.5 years. Meanwhile, the gap between what the UN considers more developed regions (Europe, North America, Australia, New Zealand and Japan) and less developed ones has decreased from 11.7 years to 9.6 years, and the difference between more

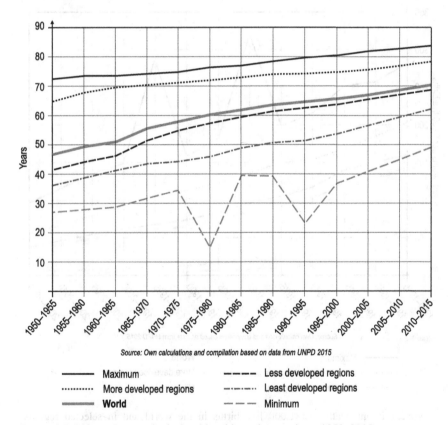

Source: Own calculations and compilation based on data from UNPD 2015

—— Maximum	– – – Less developed regions
·········· More developed regions	–··–··· Least developed regions
▬▬ **World**	– – – Minimum

Figure 2.2 Life expectancy in the world and in various regions, 1950–2015.

developed regions and the least developed countries, embracing forty-eight countries as classified by resolutions of the UN General Assembly (UNPD 2015), from 22.7 years to 16.2 years. Furthermore, this 'catching up' has taken place slightly faster in the last ten years than in the previous two decades. (It is worth noting that while the absolute minimum value for life expectancy might show remarkable swings due to, for instance, newly occurring or intensifying conflicts in certain areas, it increased by 12.5 years from Sierra Leone's 36.7 years in 1995–2000 to Swaziland's 49.2 years in 2010–2015.)

Similar patterns can be observed for infant mortality, the global value of which decreased from sixty-three per 1,000 live births in the 1990–1995 period to thirty-six between 2010 and 2015 (Figure 2.3). Here again the gap between more and less developed regions narrowed from fifty-eight to thirty-four, and from ninety-four to fifty-two between the least developed countries and more developed regions. The maximum value also declined in the same period to ninety-six in Angola and Chad, which is lower, and therefore better, than the average value for the least developed countries as a whole in as late as the first

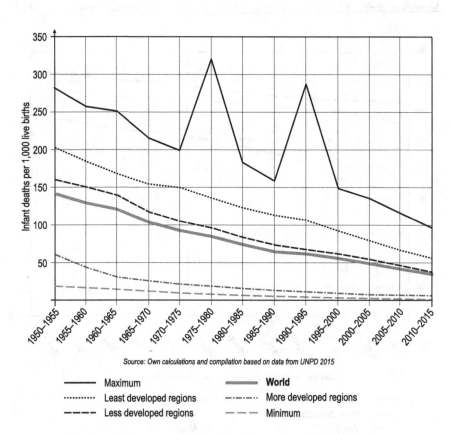

Source: Own calculations and compilation based on data from UNPD 2015

———	Maximum	▬▬▬	**World**
············	Least developed regions	—·—·—	More developed regions
— — —	Less developed regions	– – –	Minimum

Figure 2.3 Infant deaths per 1,000 live births in the world and in selected regions, 1950–2015.

half of the 1990s. Furthermore, from 2000 onwards, there has been a slight increase in the pace at which the gap has closed.

The global adult literacy rate increased from roughly 75 per cent in 1990 to more than 85 per cent in 2014 (UNESCO 2016). This means that for the world population as a whole, including children, not only the relative share but also the total number of illiterate people has declined. In 2014, a little less than 1.07 billion people were unable to read and write, which is estimated as the lowest value in absolute numbers since 1870, when the global population was only around 1.4 billion (Rozer and Ortez-Ospina 2016).

However, these global tendencies, determined on the basis of national data-sets, hide several important and challenging features of development disparity. First, declining international inequality is mostly due to the remarkable eco-nomic growth of China, which was especially high after the country's World Trade Organization accession in 2002 and before the global crisis. If the figures are recalculated for the rest of the world without China, the level of disparity shows a much smaller decline for the 2000s (Figure 2.4). Hence, while the global

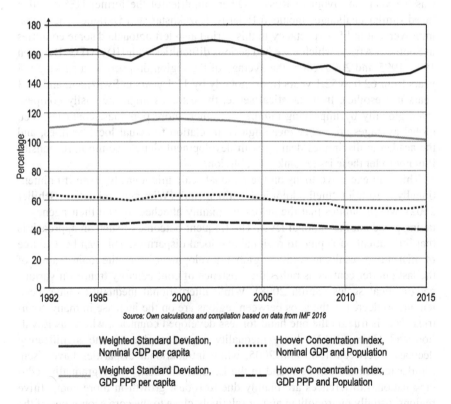

Source: Own calculations and compilation based on data from IMF 2016

———— Weighted Standard Deviation, Nominal GDP per capita

·········· Hoover Concentration Index, Nominal GDP and Population

———— Weighted Standard Deviation, GDP PPP per capita

— — — Hoover Concentration Index, GDP PPP and Population

Figure 2.4 Intercountry inequalities in per capita GDP, and the concentration of GDP and population in nominal number and purchasing power parity terms, excluding China, 1992–2015.

impact of the Chinese economic miracle should not be underestimated, the relative position of other economically less developed regions of the world has hardly improved at all in the last two decades. This also underlines the weaknesses in international development policies in the same period, which were mostly marked by the concept of structural adjustment. Therefore, new strategies are needed for the coming decades.

Second, although health and education indicators do show progress not only for China but for the least developed countries as well, there have been several countries where the local population has actually witnessed only negative tendencies. In the former Soviet Union, for example, the impact of the political and economic transition resulted in a considerable decline in life expectancy. In Russia, for instance, this value dropped from 69.1 years in 1985–1990 (for the then Soviet Socialist Republic of Russia within the USSR) to 65.0 in 2000–2005 due to increasing unemployment, economic uncertainty and the radical deterioration of the social security system. A positive trend in Russia began only with the second half of the 2000s, but the value for 2010–2015, 69.8 years, is still below the world average of 70.5 years, whereas in the late 1980s the difference was 5.5 years in favour of Russia. Other countries of the former USSR which faced similar challenges included Belarus, Kazakhstan and Ukraine. The global improvement in life expectancy in this period also left untouched some countries in Southern Africa, which have had serious difficulties with HIV/AIDS. Between 1990–1995 and 2000–2005, the average of the region dropped by a striking 9.8 years from 62.0 to 52.1 years (most notably by 13.1 years in Swaziland and 16.3 years in Lesotho). In a statistical sense, these trends might be easily compensated globally by improving numbers in other, more populous, countries. Yet, global averages are rather meaningless in relation to actual local realities, and the academic disciplines dealing with development share a common responsibility not to let these issues sink into oblivion.

Third, an exclusive focus on the national scale might easily, even if unintentionally, result in 'methodological nationalism' (Wimmer and Glick Schiller 2002), which implies that the actors are mainly attached to, and their agency is taking place on, the national level. The frequent outcome of such an approach is that little attention is paid to regional and local disparities. This can be a source of misleading results in studying global development, since the main feature of the last quarter century is rather the existence of contradictory trends on various geographical scales (Smith 2008). While international inequalities have been tending to decrease, the gaps between regions are on the increase in many countries. This is true on the one hand for less developed countries, where, as Ravallion and Chen (2012) show, inequality in per capita income significantly decreased between 1980 and 2005, while intranational disparities have risen. Similar tendencies can be found in the European Union, where gradually reducing national differences are mainly due to robust growth in more competitive regions, usually metropolitan and/or relatively close to the core economies of the EU, which in turn has led to a deepening regional gap (Puga 2002; Petrakos, Kallioras and Anagnostou 2011; Smętkowski and Wójcik 2012). Another typical example is China, where unequally allocated state investments in favour of more developed provinces and higher economic growth at the national level also massively propelled increasing regional disparities for several decades, until a gradual turn began in the 2000s (see Box 2.1).

Fourth, a rich literature exists concerning increasing socio-spatial inequality in urban contexts in the form of neighbourhood segregation and unequal access to public services and even to public spaces (Gyuris 2017). This is a common tendency in most cities of both the global North and global South, also fuelled by urban policies which have been implemented globally and adopted in very diverse local contexts in order to attract private investment to compensate for the decline in public (i.e. state) funding (Peck and Theodore 2010). Concepts like that of the creative city, for example, advertise themselves as promoting cultural heterogeneity and social tolerance, while they rather tend to benefit a highly educated and mobile global elite. This elite is becoming ever more segregated in the city environment from middle- and low-income groups, for whom labour markets are becoming very precarious and social insurance frameworks inherited from the welfare-state period increasingly eroded (Gerhard, Hoelscher and Wilson 2017; Wilson and Keil 2008; Peck 2005).

These issues, of course, do not call into question the fact that many aspects of what is often called development have indeed been improving, not only in terms of global averages, but also if the current situation in less developed regions is compared with previous decades. Between 2010 and 2015, the absolute minimum of national life expectancy values was roughly as high as the global average was in the second half of the 1950s, and life expectancy in the least developed countries was higher than the global average in the first half of the 1980s. Likewise, the globally highest value of the infant mortality rate after 2010 is nearly equal to the global average in the first half of the 1970s, while the least developed countries have the same standard as the world had roughly fifteen years ago (between 1995 and 2000), and better values than the more developed regions of the world had in the early 1950s. Therefore, global improvement in terms of many crucial indicators is a clear trend. The point might be more to underline that remarkable inequalities still exist and closing the global gap has seemingly been a less successful project over the last six decades than has improving conditions in various regions compared to their historical levels.

Beyond development? New conceptual challenges and alternative concepts for the future

After the 'golden age' of development theories and programmes between the 1950s and 1970s, from the late 1980s onwards, development has become a highly contested concept. Many do not question the relevance of the official goals of the era of development such as to increase economic production or improve life expectancy and education, but they underline that the actual policies adopted and measures taken under the aegis of development have had many controversial and even morally questionable outcomes. A frequent critique is

Box 2.1 Unequal regional distribution of investments in China: from the 'polarizing' state towards the 'equalizing' state?

A national government has some capacity to influence spatial disparities of development in its country. The strength of this influence depends on many factors, including the share of state-owned and/or state-controlled property in the economy, or the degree of centralization and authoritarianism in the political system. State intervention (of any intensity) also varies greatly as to whether it is at all aimed at reducing spatial inequalities, or is ready to sacrifice decreased disparity for other goals, such as higher economic growth on the national level. In China, which after the death of Mao Zedong in 1976 was on a par with Nepal and Bangladesh in terms of per capita GDP (Maddison-Project, 2013), the policy of 'reform and opening' launched by Deng Xiaoping in 1978 set economic growth as the ultimate goal (Wu 2008). This resulted in policies which not simply tolerated, but in some ways even increased regional disparities (Wang and Hu 1999). Gradual economic reforms focused, in the first period, on five special economic zones (Shenzhen, Zhuhai, Shantou, Xiamen and the island of Hainan) along the south-eastern coastline. In the early phase, only these areas were opened up to foreign investors. Later, they were followed by another fourteen ports in 1984 and gradually by inland areas, too (Li, Fumin and Lei 2010). Yet, the first special economic zones (SEZs) and the provinces in which they are situated have long kept their advantage, not only due to their favourable locations, but also through path dependence.

Regional disparities were also fuelled by a conscious national policy of allocating various economic tasks to different macroregions. The underlying concept was that in such a large and, at that time, poor country as China, economic productivity and the standard of living could not be lifted evenly over diverse regions and social groups, or, as Deng put it, that 'some must get rich first' (Wu 2008). Hence, to maximize national economic growth, the coastline regions were expected to attract foreign investment and knowledge and thus, on the basis of low-cost manufacturing, to produce competitive goods for the global market which could generate large revenues for the country. Central and western regions by contrast had the task of providing cheap resources for the coastline zone (Meng 2003). The national price system was designed to hold down the prices of mineral resources provided by western districts. The household registration system called *hukou*, which had kept internal migration to a minimum over the 1960s and 1970s (Cheng and Selden 1994), was loosened gradually to enable the flow of surplus cheap rural labour from the inner provinces to the coastline urban districts (Chan 2010; Csanádi, Nie and Li 2015). And, most notably, state investments along with investments from other sources – the latter through legal regulations – were predominantly channelled to the most developed eastern zones to foster even higher economic growth there, instead of being used to counteract the increasing divide between more and less developed regions.

This trend is clearly visible in the Pearson product-moment linear correlation coefficient (r) for provincial datasets of per capita GDP and per capita investment in fixed assets published by the National Bureau of Statistics of China (Figure 2.5). After 1979, for a quarter of a century, r constantly remained above the strikingly high value of +0.9, and in many years, it was above +0.95. In fact, until the late

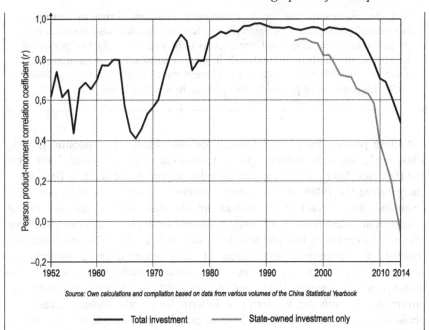

Source: Own calculations and compilation based on data from various volumes of the China Statistical Yearbook

———— Total investment   –––––– State-owned investment only

Figure 2.5 The Pearson product-moment correlation coefficient for per capita GDP versus total per capita investment (black curve), and per capita GDP versus per capita state-owned investment (grey curve).

1990s, state investments had a similarly disproportionate focus in regional terms. In 2000, the national government launched a large-scale development project to 'open up the West' (Goodman 2004), partly in order to prevent extreme regional disparities giving rise to increased political tensions. In consequence, *r* for state investments gradually decreased, but when the global crisis broke out in 2008, it was still above the quite high level of +0.6. Meanwhile, the same value for all investments exceeded +0.8, as the share of state investment in the total was declining, and non-state investors showed no inclination to give up their firm preference for more developed regions.

The global crisis, however, resulted in a thoroughgoing shift in state investments. In light of the fact that global demand for Chinese manufacturing goods declined in 2008, and significant growth seemed highly improbable for the coming years, national governments tried to counteract challenges to economic growth by allocating much more investment than before to the poorer inner provinces. These funds were mostly aimed at infrastructural development, which was expected to lay the basis for higher growth, including increased demand for goods and services in the less developed inner regions, with the ultimate goal of creating an internal market in the long run wealthy enough to buy those products which had formerly been sold abroad. Due to this change, *r* for state investment fell from above +0.6 in 2008 to below 0.0 in 2014. In fact, other sorts of investment still have a focus on more developed regions (*r*=+0.59), and the share of state investment in the national total is much smaller, 24.4 per cent in 2014, than in the late 1970s.

Still, the results show that the disproportionate allocation of state investment to more developed regions, a common practice for three decades, has come to an end. A 'polarizing' state in regional terms seems to be over in the light of post-crisis improvements. Furthermore, although it would currently be too early to speak about an 'equalizing' state (which would mean more negative values, e.g. below −0.3), this seems a real possibility for China in the next five to ten years.

that these projects have often focused too one-sidedly on economic growth, while neglecting sustainability and environmental issues (McEwan 2009; Peet and Hartwick 2009; Ziai 2006). This includes projects in the Western Bloc, especially before the 1980s, in the former Communist countries, and in the BRIC countries after the turn of the millennium. Another point of criticism is that development, interpreted in primarily economic terms, has much too often been prioritized over issues like political rights and participation. This has frequently resulted in international justification of development-oriented authoritarian regimes in the global South (such as in Chile under Pinochet or Brazil during the military dictatorship after 1964), and of paternalistic top-down methods of state governance in both the Western and Eastern Blocs, and highly exclusionary decision-making processes in the case of putative growth-inducing projects in neoliberal contexts. Some also interpret development projects in the global South, mainly launched by international organizations, as quasi-colonial attempts to sustain an uneven framework of international power relations by shaping less developed regions in line with the interests of political and economic actors in the global North. Some much discussed issues in this context are the key role of companies from wealthy countries in the implementation of development projects, the asymmetry of trade between more and less developed regions, and the maintenance of colonial modes of knowledge production and dissemination in countries of the global South, including through the formal education system (Cox 2002; McEwan 2009).

Due to these factors, in many fields of study, formerly dominant developmentalist views have been displaced by *post-developmentalist* ones (Peet and Hartwick 2009; Ziai 2006), which tend to criticize many features of the implementation of development projects, but not the main underlying assumptions of developmentalist concepts. Instead, they argue for a much broader and contextualized understanding of various societies and their needs and interests. There are also some who reject the entire notion of development, underlining that developmentalist concepts have to a certain extent brought about worse situations than those they attempted to improve. Examples of this might be the negative environmental outcomes of the 'Green Revolution' (e.g. declining biodiversity due to monocultures and the widespread use of GMOs), or water management projects (such as the construction of huge dams), toxic pollution originating from new industrial plants (e.g. chemical industries), or the systematic disqualification of local forms of knowledge, which though not necessarily modern, have nurtured much more sustainable and productive social and environmental regimes than suggested by mainstream 'Western' concepts (some examples for arid contexts can be found in Davis 2016a, 2016b).

These *anti-developmentalist* views are also subject to criticism, however, for their one-sided focus on the problems of development, which is as biased as the optimism of uncritical supporters of developmentalist concepts. In addition, longer life expectancy, lower infant mortality or higher standards of education, which are measurable outcomes of the era of development on the global stage, are phenomena which very few societies consider to be anything other than positive. Hence, many trends seem to suggest that development, both theoretically and the way it has been interpreted in practice for many decades, is too complex a concept and its actual results have been too diverse to be simply judged as 'good' or 'bad' (cf. Pingali 2012 on the outcomes of the 'Green Revolution', a cornerstone of many 'classical' development projects in the global South). Development is instead a contestable and contested but not meaningless concept, which might better be subject to permanent critical evaluation and reinterpretation according to different perspectives and the interests of various actors including the most impoverished ones. This could help identify elements that it would be reasonable to implement again in the future, as well as ones that have *ex post* turned out to be more harmful than beneficial.

In fact, more technocratic narratives of development have been to some extent influenced by the criticisms discussed in this chapter. This is reflected, for example, in the increasing attention paid by the United Nations to the Human Development Index instead of economic production only, and the UN Millennium Development Goals set in 2000 for the following fifteen years. The latter referred to economic production as only one of eight major goals, of which another explicitly related to ensuring environmental sustainability. Also among the 'MDGs' was a Global Partnership for Development within which more developed countries would contribute to a larger degree, for instance through debt relief. As the numbers presented in the Beyond Development? section of this chapter reveal, for such diverse factors as life expectancy, infant mortality, literacy and even per capita GDP, the years since the turn of the millennium have indeed brought greater improvement for less developed regions, including the least developed ones, than the 1990s. It would, however, be extremely difficult to determine to what extent this positive change can be attributed to the Millennium Development Goals, or whether it is the result of other social, economic and political shifts in the global context.

Yet, as we have seen, numbers changing for the better can, and actually do, hide remarkable inequalities within analytical categories and at lower scales. This ultimately means individual people who cannot benefit as much from the development process as is desirable. Furthermore, global challenges and the way they are perceived and conceptualized by different societies are continuously changing, partly because problems (e.g. climate change or environmental pollution), which formerly attracted little or no attention turn out to be highly important and might even reach a dangerous level due precisely to development projects which previously did not take them into consideration. Thus, a very important task both now and in the future is to utilize concepts like development, or its alternatives, not in an 'art for art's sake' way, but to adjust them again and again to changing realities and social, economic and political contexts.

Acknowledgements

This article has been supported by the National Research, Development and Innovation Office – NKFIH, under grant PD 121127. I am grateful to Ferenc Probáld for his valuable comments on earlier versions of the manuscript.

References

Agnew, J.A. (1993) 'The United States and American hegemony', in P.J. Taylor (ed.), *Political Geography of the Twentieth Century: A Global Analysis*. London: Belhaven, 207–238.

Andersen, L.E., Granger, C.W.J., Reis, E.J., Weinhold, D. and Wunder, S. (2002) *The Dynamics of Deforestation and Economic Growth in the Brazilian Amazon*. Cambridge: Cambridge University Press.

Balbi, A. and Guerry, A-M. (1829) *Statistique comparé de l'état de l'instruction et du nombre des crimes dans les divers arrondissements des Académies et des Cours Royales de France*. Paris: Jules Renouard.

Booth, C. (1902–1903) *Life and Labour of the People in London*, 3rd edn, 17 vols. London: Macmillan.

Borts, G.H. and Stein, J.L. (1964) *Economic Growth in a Free Market*. New York: Columbia University Press.

Brain, S. (2010) The Great Stalin Plan for the Transformation of Nature. *Environmental History* 15: 670–700.

Brun, E. and Hersh, J. (1990) *Soviet–Third World Relations in a Capitalist World: The Political Economy of Broken Promises*. New York: Palgrave Macmillan.

Bullard, A. (2000) *Exile to Paradise: Savagery and Civilization in Paris and the South Pacific, 1790–1900*. Stanford, CA: Stanford University Press.

Bunce, V. (1985) The Empire Strikes Back: The Evolution of the Eastern Bloc from a Soviet Asset to a Soviet Liability. *International Organization* 39: 1–46.

Butlin, R.A. (2009) *Geographies of Empire: European Empires and Colonies c. 1880–1960*. Cambridge: Cambridge University Press.

Campbell, R.W. (1985) The Conversion of National Income Data of the U.S.S.R. to Concepts of the system of National Accounts in Dollars and Estimations of Growth Rate. *World Bank Staff Working Papers Nr. 777*. Washington, DC: The World Bank.

Chan, K.W. (2010) The Household Registration System and Migrant Labor in China: Notes on a Debate. *Population and Development Review* 36: 357–364.

Cheng, T. and Selden, M. (1994) The Origins and Social Consequences of China's Hukou System. *The China Quarterly* 139: 644–668.

Conklin, A.L. (1997) *A Mission to Civilize: The Republican Idea of Empire in France and West Africa, 1895–1930*. Stanford, CA: Stanford University Press.

Cox, K.R. (2002) *Political Geography: Territory, State, and Society*. Oxford: Blackwell.

Csanádi, M., Nie, Z. and Li, S. (2015) Crisis, Stimulus Package and Migration in China. *China & World Economy* 23: 43–62.

Davis, D.K. (2016a) Political Economy, Power, and the Erasure of Pastoralist Indigenous Knowledge in the Maghreb and Afghanistan, in P. Meusburger, T. Freytag and L. Suarsana (eds), *Ethnic and Cultural Dimensions of Knowledge*. Dordrecht: Springer, 211–228.

Davis, D.K. (2016b) *The Arid Lands: History, Power, Knowledge*. Cambridge, MA: MIT Press.

Dickinson, E. (2011) GDP: A Brief History: One Stat to Rule Them All. *Foreign Policy* 3 January. Retrieved 6 October 2016 from http://foreignpolicy.com/2011/01/03/gdp-a-brief-history/.

Dickson, D.P. and Jones, J.P. (2006) Feminist Geographies of Difference, Relation, and Construction, in S. Aitken and G. Valentine (eds), *Approaches to Human Geography*. London: SAGE, 42–56.

Dorband, J. (2010) *Politics of Space: The Changing Dynamics of the 'Middle East' as a Geo-Strategic Region in American Foreign Policy*. Heidelberg: Selbstverlag des Geographischen Instituts der Universität Heidelberg.

Dupin, C. (1826) *Carte figurative de l'instruction populaire de la France*. Jobard (naïve).

Ekbladh, D. (2002) 'Mr. TVA': Grass-Roots Development, David Lilienthal, and the Rise and Fall of the Tennessee Valley Authority as a Symbol for U.S. Overseas Development, 1933–1973. *Diplomatic History* 26: 335–374.

Ekbladh, D. (2010) *The Great American Mission: Modernization and the Construction of an American World Order*. Princeton, NJ: Princeton University Press.

Fincher, R. and Iveson, K. (2012) Justice and Injustice in the City. *Geographical Research* 50: 231–241.

Friedmann, J. (1966) *Regional Development Policy: A Case Study of Venezuela*. Cambridge, MA: MIT Press.

Gerhard, U., Hoelscher, M. and Wilson, D. (eds) (2017) *Inequalities in Creative Cities: Issues, Approaches, Comparisons*. New York: Palgrave Macmillan.

Gilbert, A. and Goodman, D. (1976) Regional Income Disparities and Economic Development: A Critique, in A. Gilbert (ed.), *Development Planning and Spatial Structure*. London: Wiley, 113–141.

Glasmeier, A. (2002) One Nation, Pulling Apart: The Basis of Persistent Poverty in the USA. *Progress in Human Geography* 26: 155–173.

Goodman, D.S. (2004) The Campaign to 'Open Up the West': National, Provincial-Level and Local Perspectives. *The China Quarterly* 178: 317–334.

Guerry, A.-M. (1833) *Essai sur la statistique morale de la France*. Paris: Crochard.

Guerry, A.-M. (1864) *Statistique morale de l'Angleterre comparée avec le statistique morale de la France, d'après les comptes de l'administration de la justice criminelle en Angleterre et en France, etc.*. Paris: Baillière et fils.

Györi, R. and Gyuris, F. (2015) Knowledge and Power in Sovietized Hungarian Geography, in P. Meusburger, D. Gregory and L. Suarsana (eds), *Geographies of Knowledge and Power*. Dordrecht: Springer, 203–233.

Gyuris, F. (2014a) *The Political Discourse of Spatial Disparities: Geographical Inequalities Between Science and Propaganda*. Cham: Springer.

Gyuris, F. (2014b) Basic Education in Communist Hungary: A Commons Approach. *International Journal of the Commons* 8: 531–553.

Gyuris, F. (2017) Urban Inequality: Approaches and Narratives, in U. Gerhard, M. Hoelscher and D. Wilson (eds), *Inequalities in Creative Cities: Issues, Approaches, Comparisons*. New York: Palgrave Macmillan, 41–76.

Hanson, S. and Pratt, G. (1995) *Gender, Work, and Space*. Oxford: Routledge.

Harris, S. E. (1957) *International and Interregional Economics.* New York: McGraw-Hill.

Harvey, D. (1973) *Social Justice and the City.* London: Edward Arnold.

Harvey, D. (1978) The Urban Process under Capitalism: A Framework for Analysis. *International Journal of Urban and Regional Research* 2: 101–131.

Harvey, D. (2006) *The Limits to Capital.* London: Verso.

Harvey, D. (2007) *A Brief History of Neoliberalism.* Oxford: Oxford University Press.

Hirschman, A.O. (1958) *The Strategy of Economic Development.* New Haven, CT: Yale University Press.

Hong, Y-S. (2015) *Cold War Germany, the Third World, and the Global Humanitarian Regime.* New York: Cambridge University Press.

Huang, Y. and Leung, Y. (2009) 'Measuring Regional Inequality: A Comparison of Coefficient of Variation and Hoover Concentration Index. *The Open Geography Journal* 2: 25–34.

Hunt, A. (2002) Measuring Morals: The Beginnings of the Social Survey Movement in Canada, 1913–1917. *Social History* 35: 171–194.

Huntington, E. (1915) *Civilization and Climate.* New Haven, CT: Yale University Press.

Huntington, E. (1927) *The Human Habitat.* New York: van Nostrand.

IMF (International Monetary Fund) (2016) *World Economic Outlook Database, April 2016.* Retrieved 30 September 2016 from www.imf.org/external/pubs/ft/weo/2016/01/weodata/index.aspx.

Jefferson, M. (1911) The Culture of the Nations. *Bulletin of the American Geographical Society* 43: 241–265.

Klüter, H. (2000) 'Der Norden Russlands – vom Niedergang einer Entwicklungsregion.' *Geographische Rundschau* 52: 12–20.

Kramer, P.A. (2002) Empires, Exceptions, and Anglo-Saxons: Race and Rule Between the British and United States Empires, 1880–1910. *The Journal of American History* 88: 1315–1353.

Lenin, V.I. (1964a) 'Imperialism, the Highest Stage of Capitalism', in G. Hanna (ed.), *V. I. Lenin. Collected Works. Volume 22. December 1915–July 1916.* Moscow: Progress Publishers, 185–304.

Lenin, V.I. (1964b) The Development of Capitalism in Russia: The Process of the Formation of a Home Market for Large-Scale Industry, in *V.I. Lenin: Collected Works. Volume 3.* Moscow: Progress Publishers, 21–607.

Li, W., Fumin, S. and Lei, Z. (2010) *China's Economy.* Beijing: China Intercontinental Press.

Luxemburg, R. (2003 [1913]) *The Accumulation of Capital.* London: Routledge.

MacLeod, G. and McFarlane, C. (2014) Introduction: Grammars of Urban Injustice. *Antipode* 46: 857–873.

Maddison-Project, The (2013). Retrieved 22 September 2016 from www.ggdc.net/maddison/maddison-project/home.htm.

Marer, P. (1974) The Political Economy of Soviet Relations with Eastern Europe, in S.J. Rosen and J.R. Kurth (eds), *Testing Theories of Economic Imperialism.* Lexington, KY: D.C. Heath, 231–260.

Marer, P. (1976) Has Eastern Europe Become a Liability to the Soviet Union?, in C. Gati (ed.), *The International Politics of Eastern Europe.* New York: Praeger, 59–81.

Marx, K. and Engels, F. (1998 [1848]) *The Communist Manifesto.* London: ElecBook.

Massey, D. (1994) *Space, Place, and Gender.* Minneapolis, MN: University of Minnesota Press.

McEwan, C. (2009) *Postcolonialism and Development*. London: Routledge.

Meng, G. (2003) *The Theory and Practice of Free Economic Zones: A Case Study of Tianjin/People's Republic of China*. Frankfurt am Main: Peter Lang.

Meusburger, P. (1997) Spatial and Social Inequality in Communist Countries and in the First Period of the Transformation Process to a Market Economy: The Example of Hungary. *Geographical Review of Japan (Ser. B)* 70: 126–143.

Meusburger, P. (1998) *Bildungsgeographie: Wissen und Ausbildung in der räumlichen Dimension*. Heidelberg: Spektrum.

Meusburger, P. (2016) The School System as an Arena of Ethnic Conflicts, in P. Meusburger, T. Freytag, T. and L. Suarsana (eds), *Ethnic and Cultural Dimensions of Knowledge*. Dordrecht: Springer, 23–53.

Moran, T.P. (2005) Kuznets's Inverted U-Curve Hypothesis: The Rise, Demise, and Continued Relevance of a Socioeconomic Law. *Sociological Forum* 20: 209–244.

Myrdal, G. (1957) *Rich Lands and Poor: The Road to World Prosperity*. New York: Harper and Brothers.

Niedermaier, H. (2009) Marxistische Theorie, in G. Kneer and M. Schroer (eds), *Handbuch Soziologische Theorien*, Wiesbaden: VS Verlag für Sozialwissenschaften, 221–236.

OECD (2008) *Key Environmental Indicators*. Paris: OECD Environment Directorate.

Ostrovityanov, K.V., Shepilov, D.T., Leontyev, L.A., Laptev, I.D., Kuzminov, I.I. and Gatovksy, L.M. (eds) (1957 [1954]) *Political Economy: A Textbook Issued by the Economics Institute of the Academy of Sciences of the U.S.S.R.*. London: Lawrence & Wishart.

Parent du Châtelet, A. J-B. (1836) *De la prostitution dans la ville de Paris, considérée sous le rapport de l'hygiène publique, de la morale et de l'administration: ouvrage appuyé de documens statistiques puisés dans les archives de la Préfecture de police* (two volumes). Paris: Jean-Baptiste Baillière.

Peck, J. (2005) Struggling with the Creative Class. *International Journal of Urban and Regional Research* 29: 740–770.

Peck, J. and Theodore, N. (2010) Mobilizing Policy: Models, Methods, and Mutations. *Geoforum* 41: 169–174.

Peet, R. and Hartwick, E. (2009) *Theories of Development: Contentions, Arguments, Alternatives*. New York: The Guilford Press.

Petrakos, G., Kallioras, D. and Anagnostou, A. (2011) Regional Convergence and Growth in Europe: Understanding Patterns and Determinants. *European Urban and Regional Studies* 18: 375–391.

Pingali, P.L. (2012) Green Revolution: Impacts, Limits, and the Path Ahead. *Proceedings of the National Academy of Sciences of the United States of America* 109: 12302–12308.

Pollock, E. (2006) *Stalin and the Soviet Science Wars*. Princeton, NJ: Princeton University Press.

Pounds, N.J.G. (1963) *Political Geography*. New York: McGraw-Hill.

Puga, D. (2002) European Regional Policies in the Light of Recent Location Theories. *Journal of Economic Geography* 2: 373–406.

Rauch, T. (2009) *Entwicklungspolitik: Theorien, Strategien, Instrumente*. Braunschweig: Westermann.

Ravallion, M. and Chen, S. (2012) Monitoring Inequality. *Let's Talk Development Blog, The World Bank*. Retrieved 7 October 2016 from http://blogs.worldbank.org/developmenttalk/monitoring-inequality.

Roser, M. and Ortiz-Ospina, E. (2016) Literacy. OurWorldInData.org. Retrieved 22 September 2016 from https://ourworldindata.org/literacy/.

Rostow, W.W. (1960) *The Stages of Economic Growth: A Non-Communist Manifesto.* Cambridge: Cambridge University Press.

Schlüter, O. (1908) *Ferdinand v. Richthofen's Vorlesungen über Allgemeine Siedlungs- und Verkehrsgeographie.* Berlin: Dietrich Reimer.

Schroeder-Gudehus, B. (2014) *Les scientifiques et la paix: La communauté scientifique internationale au cours des anées 20.* Montreal: Les Presses de l'Université de Montréal.

Scott, J.C. (1998) *Seeing Like a State: How Certain Schemes to Improve the Human Condition Have Failed.* New Haven, CT: Yale University Press.

Smętkowski, M. and Wójcik, P. (2012) Regional Convergence in Central and Eastern European Countries: A Multidimensional Approach. *European Planning Studies* 20: 923–939.

Smith, N. (2008) *Uneven Development: Nature, Capital and the Production of Space*, 3rd edn. Athens, GA: University of Georgia Press.

Sneddon, C. (2015) *Concrete Revolution: Large Dams, Cold War Geopolitics, and the US Bureau of Reclamation.* Chicago, IL: University of Chicago Press.

Solarz, M. (2014) *The Language of Global Development: A Misleading Geography.* London: Routledge.

Stalin, J.V. (1972 [1952]) *Economic Problems of Socialism in the U.S.S.R.* Beijing: Foreign Languages Press.

Treaty of Peace with Germany (Treaty of Versailles) (1919). Retrieved 6 October 2016 from http://net.lib.byu.edu/~rdh7/wwi/versailles.html.

UNDP (United Nations Development Programme) (2009) *Human Development Report 2009, Overcoming Barriers: Human Mobility and Development.* Basingstoke: Palgrave Macmillan.

UNDP (United Nations Development Programme) (2010) *Human Development Report 2010, 20th Anniversary Edition: The Real Wealth of Nations: Pathways to Human Development.* Basingstoke: Palgrave Macmillan.

UNDP (United Nations Development Programme) (2015) *Human Development Report 2015, Work for Human Development.* New York: UNDP.

UNESCO (2016) *UIS.Stat.* Retrieved 22 September 2016 from http://data.uis.unesco.org/Index.aspx?DataSetCode=EDULIT_DS&popupcustomise=true&lang=en.

UNPD (United Nations Population Division) (2015) *World Population Prospects: The 2015 Revision.* Retrieved 22 September 2016 from https://esa.un.org/unpd/wpp/Download/Standard/Population/.

Wang, S. and Hu, A. (1999) *The Political Economy of Uneven Development: The Case of China.* Armonk, NY: M.E. Sharpe.

Ward, K. and England, K. (2007) Introduction: Reading Neoliberalization, in K. England and K. Ward (eds), *Neoliberalization: States, Networks, Peoples.* Malden: Blackwell, 1–22.

Williamson, J.G. (1965) 'Regional Inequality and Process of National Development: A Description of the Patterns. *Economic Development and Cultural Change* 13 4(II): 3–84.

Wilson, D. and Keil, R. (2008) The Real Creative Class. *Social & Cultural Geography* 8: 841–847.

Wimmer, A. and Glick Schiller, N. (2002) Methodological Nationalism and Beyond: Nation-State Building, Migration and the Social Sciences. *Global Networks* 2: 301–334.

Woolley, J.T. and Peters, G. (n.d.) Harry S. Truman: Inaugural Address. *The American Presidency Project*. Santa Barbara, CA. Retrieved 6 October 2016 from www. presidency.ucsb.edu/ws/print.php?pid=13282.

Wu, F. (2008) China's Great Transformation: Neoliberalization as Establishing a Market Society. *Geoforum* 39: 1093–1096.

Young, I.M. (1990) *Justice and the Politics of Difference*. Princeton, NJ: Princeton University Press.

Ziai, A. (2006) *Zwischen Global Governance und Post-Development: Entwicklungspolitik aus diskursanalytischer Perspektive*. Münster: Westfälisches Dampfboot.

3 Many worlds, one planet

Ambiguous geographies of the contemporary international community

Marcin Wojciech Solarz

Spatial development terminology: an ambiguous legacy of the past

The 1940s saw a change in thinking about development and underdevelopment, for in a period of just a few years, these issues rose to a position of central importance in international politics. This resulted in, on the one hand, attempts to classify and regionalize countries according to their level of development, and, on the other, the beginnings of an associated terminological 'Big Bang' (Solarz 2014). *Homo sapiens* is a combination of *homo categoricus* (a categorizing being) and *homo nominans* (a naming being), so when development and underdevelopment, with its clear spatial repercussions, became the object of profound attention on a global scale, it was presumably inevitable that this would trigger a process of categorizing and naming particular segments of the international community according to developmental criteria. This process of classifying and labelling, turned out to be a mad rush which combined with the relativity and malleability of the subject matter, as well as human subjectivity, meant that the hoped for well-ordered and transparent 'French garden' of categories and terms did not materialize. What emerged instead was a wild and chaotic jumble. Of these attempts to categorize the international community on the basis of divergences in development, two have proved to be of the greatest importance and durability: the division into three worlds and the dichotomous North–South.

The term 'Third World' was introduced into the language of global development in 1952, in a short article by Alfred Sauvy titled 'Trois mondes, une planète' ('Three Worlds, One Planet') published in the French weekly *L'Observateur* (Sauvy 1952). From this point on, for the next forty years until the end of the Cold War in 1989–1991, the international community was widely considered as divided into three worlds: the First made up of the world's most developed countries; the Second, encompassing the communist states, which claimed to be the First, and whose globalization would bring about the end of history (see Box 3.1); and the Third containing the entire undeveloped rest of the world (Solarz 2014). This division certainly proved to be one of the most influential in history, despite being, from the beginning, unclear in content and geographically imprecise (Figure 3.1). It can be argued that the formation of this

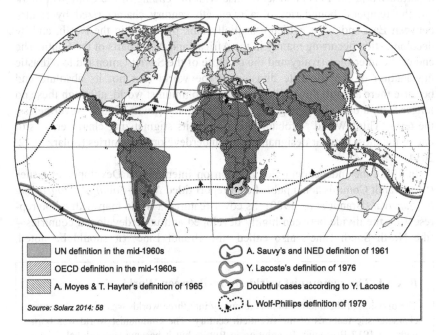

▨ UN definition in the mid-1960s	A. Sauvy's and INED definition of 1961
▨ OECD definition in the mid-1960s	Y. Lacoste's definition of 1976
▨ A. Moyes & T. Hayter's definition of 1965	Doubtful cases according to Y. Lacoste
Source: Solarz 2014: 58	L. Wolf-Phillips definition of 1979

Figure 3.1 Selected ways of delimiting 'Third World'.

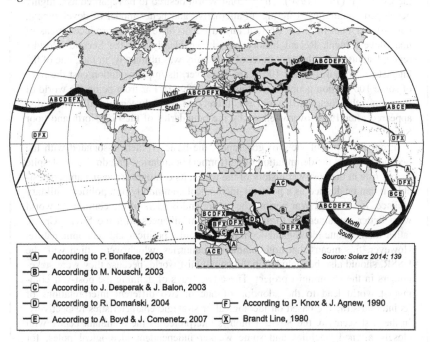

—Ⓐ— According to P. Boniface, 2003
—Ⓑ— According to M. Nouschi, 2003
—Ⓒ— According to J. Desperak & J. Balon, 2003
—Ⓓ— According to R. Domański, 2004   —Ⓕ— According to P. Knox & J. Agnew, 1990
—Ⓔ— According to A. Boyd & J. Comenetz, 2007   —Ⓧ— Brandt Line, 1980

Source: Solarz 2014: 139

Figure 3.2 Selected ways of delimiting 'North–South'.

tripartite world (in 1917) and its demise with the ending of the Cold War mark out the temporal boundaries of a twentieth century characterized by rivalry between East and West, which essentially took place within the North, and by decolonization occurring mainly in the African and Asian parts of the South. The end of the inter-bloc rivalry and the collapse of the Soviet Union led to a drastic devaluation of the tripartite division of the world (but not its absolute end because there are still powerful relics of the communist world, although these no longer form a bloc competing for world supremacy [Swift 2003]). Nevertheless, the term 'Third World', probably because of its stigmatizing character, is still one of the most popular designations for underdeveloped countries (Solarz 2012, 2014).

In 1980, the Independent Commission on International Development Issues ('the Brandt Commission') published a report titled *North–South: A Programme for Survival* (Brandt *et al.* 1980). Its cover featured a world map with a line representing the divide between highly developed and underdeveloped countries – 'the North–South Line', also called 'the Brandt Line' (it should be noted,

---

Box 3.1 The communist world: an In-Between World

The division of the international community into three worlds was one of the most characteristic features of the twentieth century. The communist world was born with the 1917 Bolshevik Revolution in Russia but it became truly global during the Cold War (1945–1990). The Second World desired to be regarded as a highly developed community, offering an alternative and better model of development, but no later than in 1989–1991, it turned out to have merely produced areas of underdevelopment. Recent history has shown clearly that the Second World's reputed status as a pillar and driver of modernity was ultimately, in fact, one of the biggest illusions of the last 100 years. However, its implementation cost the lives of some 100 million people (Courtois *et al.* 1999: 4). Far from being the world of the future, the communist world was merely an 'In-Between World', which disappeared after 1989, overwhelmed by the boundaries of the rich North and poor South.

From the point of view of geography, the basic problem associated with the Second World is the delimitation of its (former) boundaries. Paradoxically, despite its key role in the post-war history of the world, its contours are vague and contestable. It can be defined as a political–ideological community or a political–military alliance. In the Cold War period, the nature and borders of the communist world were better explained by formal and informal alliances with the Soviet Union, rather than opaque ideological–legal indicators of a political and economic nature. However, this means that non-communist countries, which were allied with the USSR, should also be included in the communist world, because they were stakeholders in the communist project. These two ways of defining the post-war communist world lead to the adoption of one of two different interpretations of its internal structure based on ideological or strategic characteristics respectively. In the first scenario, the Soviet Union is interpreted as the dominant centre with closely aligned satellites and some weaker independent ideological poles, for example, China and Yugoslavia (after their break with the USSR). In the second, the Soviet Union is the core of the international communist system and all countries within it are satellites, but with different degrees of dependence and autonomy. A separate unresolved problem is whether or not a 'remnant' communist world actually exists today, and, if one does, what its boundaries are (Figures 3.3 and 3.4 (see more Solarz 2014: 28–41).

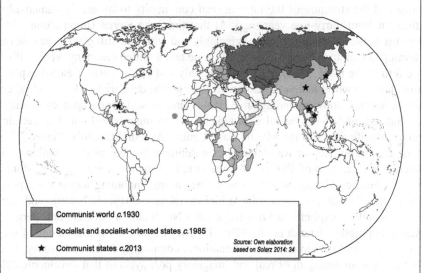

Figure 3.3 The dual nature of the communist world: the communist world as a political–ideological community.

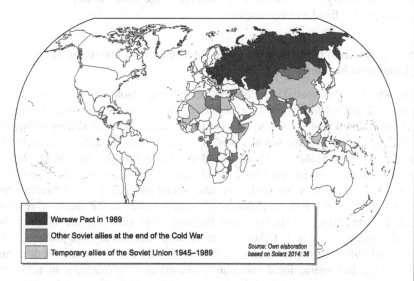

Figure 3.4 The dual nature of the communist world: the communist world as a political–military alliance.

however, that the beginnings of thought about the world through the prism of a North–South dichotomy reach as far back as the late 1950s). One of the most amazing aspects of the Brandt Line is the extent to which it has enslaved our imagination and continues to greatly influence our picture of the international community. It seems to make no difference that the shape of international relations and the structure of the international community today are fundamentally different from thirty-five years ago. At that time, the Soviet Union along with the rest of the communist bloc convincingly and with flair still played the role of developed countries, post-Maastricht Europe and a united Germany were still in the future, the label 'Made in China' was only just on the cusp of global expansion, and the world had yet to experience the global debt crisis. These and other dissimilarities notwithstanding, the Brandt line is still reproduced essentially unchanged, and if it is at all adjusted, this is on the whole limited to certain minor modifications in the Middle East, Central Asia and Oceania (Figure 3.2). That the Brandt Report was imprecise in defining the line of demarcation and unclear in its criteria of division also seems to be a matter of indifference. Moreover, it has been forgotten that from the beginning, including within the report itself, the Brandt Line met with criticism (Solarz 2014: 127–139), and the report's authors explicitly acknowledged that North and South were not 'permanent grouping[s]' (Brandt *et al*. 1980: 31). The Brandt Line is indeed one of the most influential divisions of the international community in human history, but in fact, it is an amalgam of real and imaginary portrayals of that developmental divide. With the end of the Cold War, it should have been consigned to the history books. Development is a process, as are its individual stages, and therefore no boundary drawn between the developed and underdeveloped worlds can be accepted as fixed and unchanging.

A hypothetical evolutionary cycle for the spatial language of global development

The Big Bang of spatial development terminology took place over roughly a forty-year period following the outbreak of the Second World War (i.e. up to *c*.1980). This period witnessed the rise and proliferation of a number of hugely influential terms for 'the world's worlds' distinguished by differences in levels of development achieved. These included: 'Third World', 'developing countries', 'least developed countries', 'South' and 'North'. The terminological Big Bang was triggered by political, economic, social and technological shocks connected with rapid world development ultimately rooted in the industrial revolution (Solarz 2014, 2017). The years between 1941 (the first use of the term 'underdeveloped areas' in today's understanding occurred in 1942 [Solarz 2014: 50]) and 1980 (the popularization of 'North–South', the last great pair of terms describing the international community according to differences in achieved levels of development) saw the birth and spread of all the key terms used today to describe the world's developmentally differentiated structure. Since 1980, there has been no instance of the emergence and popularization of a new term for developmental differentiation on a global scale, which could compete in popularity with the old terminology. Some success has been experienced only by some new terms for smaller segments of the international community, which tend to refer to involvement in development processes (in its various aspects, e.g. 'emerging markets', 'countries in transition') and its future effects (e.g. BRIC), rather than to levels of development achieved.

The post-war terminological Big Bang is generally ascribed to the impact of four factors in particular which began to affect international relations from the last years of the first half of the twentieth century and led to issues of development and underdevelopment taking centre stage in the attention of the international community from the 1940s onwards. These factors were the Great Depression (1929–1933), the Second World War (1939–1945), the Afro-Asian post-war wave of decolonization (especially 1945–1975) and the Cold War (1945–1989/1991) (Solarz 2014). It seems, however, that this list is incomplete, for in the mid-twentieth century, the international system and world public opinion were under extreme pressure for a number of other reasons. These included a worldwide (including in the global South) population explosion, (especially after 1927 when the world's population exceeded 2 billion), a sustained and rapid increase in global wealth, and at the same time sharply growing differences between the rich North and the poor South (increasingly spectacular in degree from the 1940s onwards, and increasingly visible thanks to the development of modern communication technology and passenger transport). There was also the simple fact of the globalization-induced 'flattening' of the world community, already radical in the early twentieth century (Solarz 2017), and taken much further by the development of electronic media in the second half of the twentieth century. In the mid-twentieth century, the international community thus found itself in an unprecedented situation – never had so many people been so close to each other, divided between some countries which were spectacularly growing and the underdeveloped remainder, ready (and able) even to annihilate the whole of human civilization. Development and underdevelopment ceased to be mere passive descriptions of reality, but became challenges, also at an ethical level, and important factors in international politics (the power games of the superpowers included). Synergies between the various pressures on the international system referred to above had led to a revolution in the awareness of world public opinion and politicians, resulting in development and underdevelopment entering the category of key global issues, and this in turn provided a strong impetus to create names for the world's worlds which would indicate their levels of development.

The birth of this specific type of geographical name can be treated as an historically isolated coincidence, the result of particular circumstances existing in the mid-twentieth century, and this is how the matter is generally viewed today. However, a tempting hypothesis can also be advanced that these terms were produced by a cyclical evolutionary process, which began in the eighteenth century (Figure 3.5). This hypothetical historical cycle has no past precedent: it is the first, is still ongoing and it is not known whether it will be repeated in the

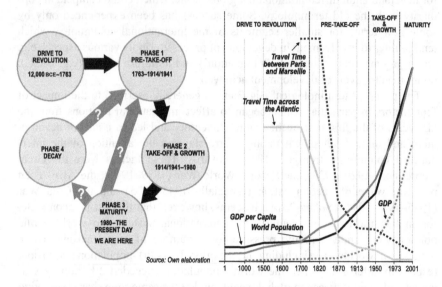

Figure 3.5 A hypothetical evolutionary cycle for the spatial language of global
development.

future (it may be a one-off cycle, since a future event inducing a new cycle may
not occur). This undoubtedly makes the hypothesis weaker.

The model proposed here for the evolution of the language of global development consists of four phases, preceded by an initial phase which the author is
calling *drive to revolution*.[1] This phase encompasses the period in which global
divisions of an objective character have already emerged, shaped by the Agricultural Revolution and subsequent development processes. However, at that time,
there clearly existed no global, common language to designate societies divided
into groups according to differences in levels of development. A cardinal feature
of the then world was the profound mutual isolation of its segments (which only
began to break down with the advent of the Age of Discovery in the fifteenth
century), although local/regional centres of development did try to name the
parties to local/regional developmental rifts (e.g. the Greeks and Romans versus
barbarians). In this phase of the cycle, the world was changing, and poles of
development did emerge, but from the perspective of the coming centuries these
changes were not spectacular. The world was characterized by continuity rather
than change. This period lasted from about 12,000 BC (the beginning of the Agricultural Revolution) to about AD 1763 (the symbolic beginning of the Industrial
Revolution with James Watt's enhancement of the steam engine).

The first phase of the evolutionary cycle of the language of global
development – *pre-take-off* – was opened by a great revolution. The Industrial
Revolution, when 'the world had slipped its moorings' (Landes 1998: 192),
brought about colossal changes, increasing by degrees and cumulative, in the

entire international system including its structure. During this period (about 1763–1914/1941), divisions within the international community were constantly deepening, but there was as yet no sense of need to name the newly observed divisions.

> It presumably could not have been otherwise in a period when many eminent scholars claimed that the key global process was the progressive Europeanization of the world (Pawłowski 1938). From this perspective, and from the point of view of observers who primarily represented European civilization, careful analysis of the underdeveloped rest of the world and the invention of special names for it were no doubt simply redundant since that world would inevitably at some point merge into the world of 'global Europe'. This situation only changed when, from the 1940s onwards, the process of the Europeanization of the world was called into question by international reality and public opinion, and the circle of those participating in discourse on the development and structure of the world began to continuously diversify. Language was now ready for 'toponyms' designating the world's worlds on the basis of achieved levels of development.
>
> (Solarz 2014: 49)

And so began the second phase of the cycle – *take-off and growth*. The changes taking place within the international community reached a spectacular degree and continued to rapidly gain pace. This dynamic growth and change was accompanied by the appearance and cumulation of great political, social and economic tensions which in this phase of the cycle (1914/1941–1980) were manifested in the form of large shocks/international crises – the Great Depression, world wars and conflicts (the First and Second World Wars, the Cold War) and drastic changes in the international system (the final decline in world significance of the European powers, the polarization of international relations and decolonization). From these factors, there grew a global awareness of the reversibility and instability of developmental advance and the existence of significant differences in levels of development on an international scale. It was this phase that saw the Big Bang of spatial development terminology.

Currently (from around 1980), we are in the third phase of the cycle – *maturity*, in which the structure of the world and the language describing it is stabilizing. Changes are still underway (and can still be significant and dynamic), but they have lost their spectacular character, because there is no longer a contrast between the contemporary period and the immediately preceding one. For in both of these, changes are characterized by a high level of dynamism and are, therefore, harder to discern and have less power to shock. People have become accustomed to great spatial development disparities, so their potential continuance or growth is not something which draws unusual attention and demands new terminology. Besides, there are already plenty of currently available terms developed in the previous phase of the cycle to designate groups of countries with differing levels of development which fully meet the needs of modern *homo nominans*. Hence, the fading away of the terminological Big Bang. Significant changes in the structure of the international community have already taken place and been named. As a result, only concepts of limited spatial extent are now appearing in the language of global development.

We do not know what awaits us in the future. Perhaps the *maturity* phase will last permanently. Perhaps, however, there will be another great economic revolution akin to the Industrial Revolution, which will open a new cycle (with a new *pre-take-off* phase). A new revolution will not necessarily destroy the old terminology, which could be used to describe the new divisions. However, if development processes lead to the disappearance of global spatial inequalities, then the currently used spatial development terms will go out of use or change their meanings. This would mean the start of the next, fourth phase of the cycle – *decay*, which would either finally close the whole cycle once its terms take on a historical character or different meanings, or, if it is interrupted by a new great revolution, give way to a new *pre-take-off*.

Spatial development terms: an attempted systematization

Spatial terms used in the language of global development (e.g. Third World, North, South) are not only concepts of a social, economic and political character, but also special kinds of geographical names. These concepts not only attribute certain social, economic and political features to the fragments of reality they name, but they also locate them in space, assigning them a place and boundaries. Of course, their geographical dimension is dynamic and subject to debate, but this does not change the fact that geography is inextricably linked to these concepts.

Names for the world's worlds used in the language of global development can be split into three basic groups (Figure 3.6):

- *large terms* (Third World, developing countries, South, North, etc.): these refer to large groups of countries resulting from the disjunctive division of the international community as a whole; in the case of these terms, there is, essentially, consensus only as to their very general meaning; their geography is imprecise and variable; their birth and popularization occurred in the three or four decades immediately following the Second World War; the last of them ('North' and 'South') came into widespread use in the early 1980s.
- *medium terms* (e.g. least developed countries [LDCs], landlocked developing countries [LLDCs], small island developing states [SIDS], most seriously affected states [MSAs], capital-surplus oil exporters, newly industrializing/ industrialized countries [NICs], heavily indebted poor countries [HIPCs], emerging countries/markets [Ems], countries in transition/nations in transit): these terms specify groups of up to several dozen countries distinguished from the international community as a whole. Thus, unlike the large terms, they are not the product of an exhaustive and disjunctive division of the

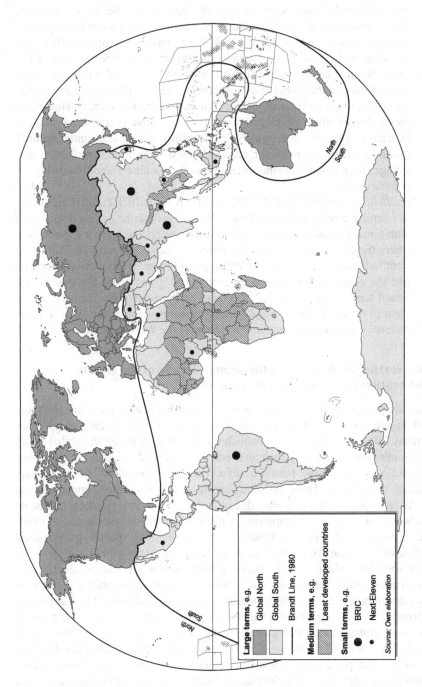

Figure 3.6 Large, medium and small spatial terms – the geographical dimension of sample terms.

whole, instead they represent smaller categories within it. The criteria and geographical boundaries represented by these terms should be subject to critical analysis and possibly adjusted as a result. It seems that the heyday of 'medium terms', perhaps still ongoing, began on the cusp of the 1970s. Two events seem to have been especially important in the development of this terminological category: first, the oil crisis which created a deep and open segmentation in the global South, and then in the late 1980s and early 1990s, the crisis in, and collapse of, the Eastern bloc which successfully challenged the myth of a single developed North (the 'countries in transition'/'nations in transit' category was widely used at that time), and simultaneously opened up new prospects for economic expansion (thus contributing to a change in meaning of the term 'emerging markets' and assisting its popularization).

- *small terms* (BRIC, BRICS, BRICM, BRICA, BRICET, BRICK, BRICI and similar acronyms, Next-Eleven, etc.): these describe groups of countries with a small number of members (between several and up to a dozen or so), where the composition of member states is precise and geographically constant, although they are also open to question due to the particular criteria and the actual intentions and content of the given term; the time of the 'small terms' came with the beginning of the twenty-first century – they seem to be an attempt to negate the dominant position of the West in international relations, or an attempt to anticipate the West's decline.

The North–South divide in the second decade of the twenty-first century: in search of a new paradigm

'Ask simple questions, because the answer to complicated questions probably will be too complicated to test and, even worse, too fascinating to give up' (Crosby 2015: 6). Attempts to establish the boundary between the global North and South lead to precisely such answers – complicated, ambiguous, multivariant and very subjective (and for these reasons, contentious and ultimately eluding consensus). These answers are, however, necessary, important and useful contributions to our knowledge of the modern world. For there can be no doubt that the world is developmentally diverse and that this fact is crucial to our efforts to describe and comprehend it. Nevertheless, a specific geographical picture of this diversity is based on many variables – the definition of development we adopt, along with its philosophical foundations, our political views, experiences and personal beliefs, education, geographical location etc. If, therefore, development can be simultaneously understood and defined in many ways, determined and measured using various criteria and indicators, then it is possible for many global Norths and global Souths to coexist at the same time and in parallel (see Box 3.2). This means that there is no single universal dividing line between developed and underdeveloped countries. We should not, however, conclude that all proposed delimitations of North and South represent reality, and among those that do, that all are equally accurate and valuable. We are faced

Box 3.2   Social contract theory and the North–South divide in the second decade of the twenty-first century

The question of development is, in fact, also addressed by social contract theories, including the most famous of these formulated by Thomas Hobbes, John Locke and Jean-Jacques Rousseau. The reason people organized themselves into states was because they wanted to resolve the problems of the state of nature. From this perspective, the development process may be seen as a sliding scale, where the starting point is the state of nature with its problems, and the end point is the political body of the state which has wholly eliminated the problems of inter-human relationships arising from the state of nature. Thus, the state of nature, along with political states based on a social contract but entirely or substantially failing the expectations of those who entered into the contract, correspond to the undeveloped and underdeveloped worlds. Political organisms which are most effective, or at least the most effective in a given era, in removing the defects of the state of nature, are situated at the opposite pole of development, constituting the highly developed world.

Hobbes, Locke and Rousseau had differing perceptions of the state of nature and, in consequence, different expectations as to the outcome of the change they described. Due to varying conceptions of the social contract, there will be varying definitions of developed/underdeveloped countries and thus various geographical depictions of the North and South.

Thomas Hobbes defined the state of nature as complete 'lawlessness' characterized by universal war (*bellum omnium contra omnes*) (Porębski 2004). Permanent insecurity and the pervasive threat of death was so unbearable that people brought the state into being. Thus, we can deem the terms of the social contract to be met when the population feels safe. The reality and perception of a society's safety and security can be measured in many ways. For the purpose of this box, two indicators are used. The first, subjective, measures the sense of security felt by a society's population, that is, those who have entered into a social contract. The second, objective, measures the number of homicides per 100,000 population. For the first indicator, we deem the terms of the social contract to have been met when more than half the population feels safe. For the second, we deem the terms of the social contract to have been met when the number of homicides is negligible, that is, less than one per 100,000 population. For the sake of simplicity and seeking to present a clean dichotomy, we also deem countries for which this data is unavailable to have not met the requirements of the social contract. The combination of these two (or actually three) criteria and the thresholds we have fixed for them enables us to determine the division of the contemporary international community between the global North and the global South from the perspective of Hobbes. Accordingly, the global North consists of eighteen geopolitical units (including seventeen countries). The rest is the global South (Figure 3.7).

John Locke in turn defined the state of nature as a state where people were exposed to injustice. So, from Locke's viewpoint, the degree of confidence in the functioning of the government and judiciary seems key to assessing the success of the social contract. In order to trace the North–South divide from Locke's perspective, we will again take into account two, or de facto three

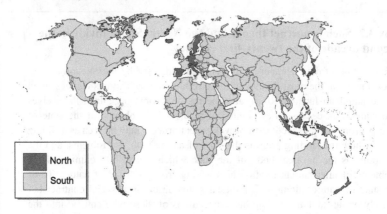

Figure 3.7 North and South in 2016 from different philosophical perspectives: the 2016 North and South in Hobbes' perspective.

criteria. This time both of the main indicators are subjective. The first is trust in the national government, while the second is confidence in the judicial system. In both cases, we deem that the objective of the social contract has been fulfilled when more than half the population has confidence in both the government and the system of justice. Once again, the third additional criterion is the presence or absence of relevant data. An absence of data again locates a given country within the boundaries of the global South. According to the criteria adopted here, the Lockean global North is made up of thirty-six very different countries. The remainder of the international community constitutes the global South (Figure 3.8).

According to Jean-Jacques Rousseau, by contrast, the state of nature had an almost paradise-like character and human life at that time was both free and happy. Unfortunately, human vice destroyed the original state of humanity and the only recourse was to enter into a social contract (Porębski 2004). Thus, the global North is made up of societies which are both free and where people are satisfied with

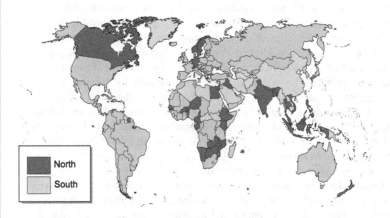

Figure 3.8 North and South in 2016 from different philosophical perspectives: the 2016 North and South in Locke's perspective.

their lives. Here also we have adopted two, or de facto three, criteria as once again the absence of data automatically assigns a given country to the category of global South. The first criterion, based on expert opinion, relates to the observance of political rights and civil liberties and thus to freedom. One organization which assesses this criterion is Freedom House, an NGO, whose Freedom in the World index is used in this analysis. To be considered here as a potential member of the global North, a country must be assessed by Freedom House as having the status of 'free' (other possible statuses – 'partly free' and 'not free' – locate a given country within the boundaries of the global South). The second criterion is subjective and relates to the perception of life satisfaction (Overall life satisfaction index). In light of this criterion, the global North is not only made up of free societies, but also those in which, on a scale of zero to ten, overall life satisfaction is evaluated by the population at more than 5.0 points, that is, it is in the upper half of the scale. The global North from the perspective of Jean-Jacques Rousseau is the largest North we have examined thus far in this box, including no fewer than fifty countries. The world's remaining countries belong to the global South (Figure 3.9).

Due to varying conceptions of the social contract, there are various Norths and Souths. Three different 'philosophies of development' have led us to three distinct pictures of the North–South divide. The pictures of the divide in the international community between highly and poorly developed countries corresponding to Hobbes', Locke's and Rousseau's philosophies are both surprising and very debatable. Which of the pictures is true – all of them, some or perhaps none?

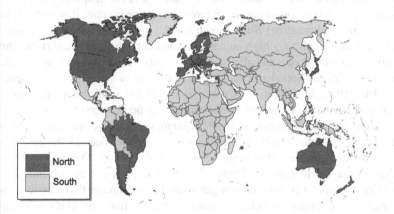

North

South

Figure 3.9 North and South in 2016 from different philosophical perspectives: the 2016 North and South in Rousseau's perspective.

with a paradox: the North–South divide does exist objectively, it is a fundamental fact affecting the entire international community, and yet there is such deep disagreement as to its nature and boundaries, that to define, identify and name it with any degree of unanimity has proven impossible. Furthermore, the answers that we do find to some of these questions only retain their validity for a given time, for development is a process. It is worth noting that this inability to build a global consensus regarding the definition and delimitation of the

North–South divide is another factor in the sustained appeal of the Brandt Line (as is the schematic and conservative way some academics and journalists approach development issues). The Brandt Line can be seen as a kind of global compromise which allows the world's developmental rift to be quickly and unambiguously identified, without the need for serious research, even though the injured party in this approach is truth itself. The East–West axis was a simple, very utilitarian, but also accurate description of international reality until the end of the 1980s (a clear us versus them divide), and its passing by no means meant that the need for a simple description of an increasingly complex world shared the same fate. North–South seems to be the successor to East–West as the world's primary structural axis, and, therefore, the question of its boundaries is one of the most important issues in contemporary international relations and human geography.

In general, discussion of the division of the international community into developed and undeveloped countries is burdened by schematism. The subject is viewed through the prism of the collocations 'rich North' versus 'poor South'. The dividing line is assumed to run between monetary wealth and poverty. Fortunately, development is no longer understood and measured only in narrow economic terms. A rich society is now considered one in which people have the widest range of opportunities. This in turn is based on people's available financial resources, their level of health and education, and the scope of guaranteed political rights and civil liberties. By reference to these criteria, it is possible to construct a new (alternative) North–South divide, in which the North is constituted by countries with very high human development, where political rights and civil liberties are observed, and all remaining countries are assigned to the South. Such a newly defined boundary could aspire to the role of being a new and better Brandt Line (though, in fact, it is not a single global continuous line, since, according to this view, the international community breaks down into islands of development and underdevelopment). This boundary, like its predecessor, also divides up international reality, but it does so in a superior fashion, because it is based on defined criteria and indicators (including the Human Development and Freedom in the World indices) (Figures 3.10 and 3.11). It has the advantage of objectivity, because it is difficult to argue with the assumption that people want to be richer, live longer, be better educated and more free, and that those who live in countries which have maximized the achievement of these goals, can be said to live in the most highly developed societies of their time. There is a problem, however, in that these categories are not as obvious and indisputable as they might seem. Wealth, longevity, extensive knowledge of the world and unfettered freedom also have their dark sides. Long life brings with it such experiences as the ageing process and old age itself, and therefore frequently also illness, infirmity and suffering. Knowledge concerning, for example, societal mechanisms and interpersonal relationships can give rise to deep frustrations. Furthermore, the absolute maximization of wealth and freedom (and the absolutization of these as social values) has unequivocally negative consequences in the form of greed and envy, excess and lack of sustainability, and

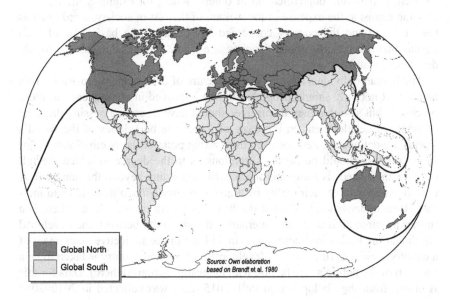

Figure 3.10 The North–South divide in 2016 according to traditional paradigms: the 'classic' Brandt Line, 1980.

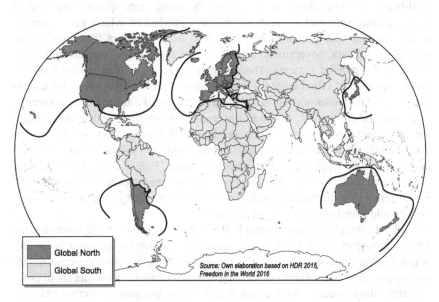

Figure 3.11 The North–South divide in 2016 according to traditional paradigms: an alternative North–South line (lines), 2016.

the enslavement and objectification of others. When, for example, the freedom of some comes at the expense of the welfare of the rest of society, a red line has been crossed and the country concerned has betrayed the basic idea of high development, that is, the good of society as a whole, and has thus lost its 'highly developed' status.

The human individual should be at the centre of development processes. This means that not only should people's lives be improved in consequence of these processes, but also that they themselves should have a say in assessing the level of achieved development on the basis of which the boundaries of the world's worlds are delineated. Perhaps even the starting point of the North–South delineation process should be people's opinions as to the degree to which the full range of their needs is being met (necessarily extending beyond the material). A candid assessment of their reality by a given country's inhabitants de facto indicates their understanding of what (high) development means. Hence, there is a further possible division of the international community between the developed North and the backward South, this time in light of the subjective assessment of a country's citizens (the level of development of a given country is taken to be a function of the public's satisfaction with the results achieved there). Data for this is drawn from the Gallup World Poll, 2015 (data was collected in 2010–2014 with results from most of the world's countries; Human Development Report 2015: 266–269). Partial scores (nine in total) relate to perceptions of individual well-being (six indicators: education quality, health care quality, standard of living, feeling safe and freedom of choice – the last of which gives separate results for men and women) and perceptions of government (three indicators: trust in the national government, actions to preserve the environment and confidence in the judicial system). The arithmetical mean of nine partial scores is taken as a country's final score for this indicator. This is a comprehensive assessment as it takes into account a full picture of social life: material resources, education, health, safety, freedom, government efficiency, justice and concern for the environment. In this perspective, the global North is made up of countries with the highest degree of public satisfaction. However, this subjective indicator and a high quality of life in the objective sense (Figure 3.12) do not always overlap, and for this reason additional country subcategories have been identified by reference to quality of life as defined above in the light of the Human Development and Freedom in the World indices.

Full data from the Gallup World Poll, 2015 is available for 158 countries and territories. Apart from thirty-seven countries and territories (usually small), which Gallup did not include in the survey, we have incomplete responses for twenty-six. The vast majority of missing answers relate to questions about political and safety issues,[2] which raises the suspicion that some governments have no wish to disclose to world public opinion their citizens' views on politically sensitive subjects. The absence of these 'sensitive scores' presumably leads to an overstatement of a country's total score, and therefore to a better position in the overall ranking (sometimes it seems even considerably better as in the case of the rich Arab states of the Persian Gulf). Therefore, in my listing the last score is

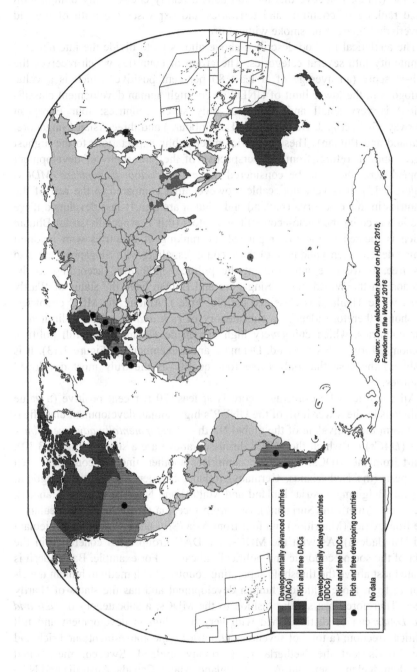

Figure 3.12 The North (DACs) – South (DDCs) divide in 2016 according to a new paradigm.

Developmentally advanced countries (DACs)

Rich and free DACs

Developmentally delayed countries (DDCs)

Rich and free DDCs

Rich and free developing countries

No data

Source: Own elaboration based on HDR 2015,
Freedom in the World 2016

always the quotient of the sum of the available answers, and the number of all possible (nine). I believe this method adds a reality check to my opinion-poll based ranking of countries and territories and expresses the truth of the old proverb that 'there is no smoke without fire'.

The analytical method described above allows us to divide the international community into several categories. The group of countries which receives the highest score (an average of at least 80 per cent positive, which is a value analogous to the lower limit of UNDP's very high human development classification) is very small and comprises only seven countries: four European (Norway, Switzerland, Denmark and Luxembourg) and three Asian (Singapore, Thailand and Bhutan). These countries meet public expectations to the highest degree and, therefore, from the perspective of the philosophy of development adopted here, they can be considered the *most developed countries (MDCs)* (Figure 3.13). It is very noticeable, however, that compared to the rest of the countries in this category, Thailand and Bhutan are clearly less developed if we take into account the socio-economic indicators (this is especially so for Bhutan which is located in the lower part of the ranking of countries with medium human development) and both of these countries, along with Singapore, are also less free, if we take into account the political indicators (according to the Freedom in the World index, Singapore and Bhutan have the status of 'Partly Free', while Thailand is classified as 'Not Free'). Among the MDC countries, we should therefore identify a subcategory of *rich and free MDCs*, limited to four countries, which enjoy very high human development and full political freedom (Norway, Switzerland, Denmark and Luxembourg) (Figure 3.13). It is doubtless the case that only these four qualify as the truly most developed countries.

All countries whose average score is at least 70 per cent positive (a value analogous to the lower limit of the UNDP's high human development classification) form the equivalent of the global North – *developmentally advanced countries (DACs)* (of which the MDCs discussed above are a subset) (Figure 3.12). Apart from the MDCs, this group comprises another nineteen countries – ten European (the Netherlands, Germany, Ireland, Sweden, the United Kingdom, Iceland, Belgium, Austria, Finland and Malta), one North American (Canada), one South American (Suriname), two from Oceania (Australia, New Zealand), one from Africa (Mauritius) and four from Asia (Sri Lanka, Philippines, Vietnam and Bangladesh). As with the MDCs, the DAC group lacks coherence in the light of the socio-economic and political indicators. For example, Bangladesh is located just above the threshold separating countries with medium human development from those with low human development and has the status of 'Partly Free'. Therefore, in a similar fashion to the MDCs, a subcategory of *rich and free DACs* can be distinguished with very high human development and full political freedom (a total of seventeen countries – the above-mentioned rich and free MDCs and the Netherlands, Germany, Ireland, Sweden, the United Kingdom, Iceland, Belgium, Austria, Finland, Malta, Canada, Australia and New Zealand) (Figure 3.12). Perhaps these are the countries, which make up the true

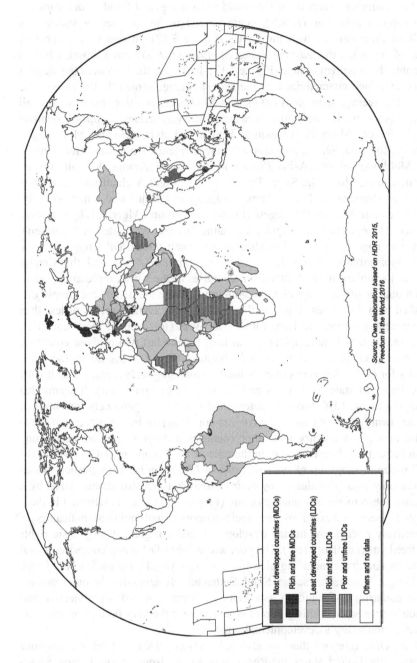

Figure 3.13 The most and least developed countries in 2016 according to a new paradigm.

Source: Own elaboration based on HDR 2015,
Freedom in the World 2016

Most developed countries (MDCs)

Rich and free MDCs

Least developed countries (LDCs)

Rich and free LDCs

Poor and unfree LDCs

Others and no data

global North, where a high quality of life is accompanied by a high degree of public satisfaction with the surrounding reality.

The remaining countries of the world form the global South – *developmentally delayed countries* (*DDCs*), characterized by lower average satisfaction levels of their citizens than the DACs (Figure 3.12). In this very incoherent group of countries, the *least developed countries* (*LDCs*) form a subset, which is undoubtedly important to note (Figure 3.13), located at the bottom of the ranking of countries and characterized by the lowest average scores. In the case of the LDCs, the average score does not exceed 50 per cent positive (but in fact, in all cases, it is lower). In this group, there are as many as sixteen European countries (Italy, Greece, Slovakia, Lithuania, Hungary, Latvia, Croatia, Montenegro, Belarus, Romania, Serbia, Bulgaria, Ukraine, Armenia, Bosnia and Herzegovina and Moldova), eleven Asian (South Korea, Saudi Arabia, Kuwait, Oman, Lebanon, Iran, Mongolia, Syria, Pakistan, Yemen and Afghanistan), five Latin American (Venezuela, Brazil, Peru, Paraguay and Haiti) and, unsurprisingly, African countries form the largest (twenty-six) group (Algeria, Libya, Gabon, Morocco, Congo, Ghana, Angola, Tanzania, Nigeria, Madagascar, Mauritania, Comoros, Togo, Swaziland, Lesotho, Sudan, South Sudan, DR Congo, Liberia, Mali, Mozambique, Sierra Leone, Guinea, Burundi, Chad and the Central African Republic). In total, fifty-eight countries of the world, representing more than a quarter of the international community, where the people are simply dissatisfied with their lives. Worst in the ranking, with average scores of less than 30 per cent positive, are Iran, Libya and Syria (for each of these countries, however, the data is incomplete, yet at least for the last two of these countries, the result does seem clearly credible because their citizens do in fact have reasons for evaluating very critically their surrounding reality, marked as it is by the collapse of state institutions and the ravages of brutal war). It seems that belonging to the LDC group is determined either by a genuinely low level of development (as in the case of the African countries), or by the perception of the public in a given country that other countries in their region are performing much better than their own (e.g. in the case of the European LDCs). Within the LDC group, a subgroup of *poor and unfree LDCs* can be distinguished, which are not only poorly evaluated by their citizens but are also characterized by a sustained objectively low quality of life (Figure 3.13). This is reflected in these countries being assessed by the socio-economic and political indicators as respectively having low human development and not guaranteeing basic rights and freedoms (i.e. 'Not Free'). The poor and unfree LDC group comprises a total of eleven countries, including nine from Africa (Angola, Burundi, CAR, Chad, DR Congo, Mauritania, Sudan, South Sudan and Swaziland). The other two are Asian countries (Afghanistan and Yemen). According to both subjective and objective assessments, this group of countries is at the very bottom of the international community's developmental scale.

In the DDC category, there are also *rich and free DDCs* – a total of twenty-one countries (the United States, the Republic of Korea, Israel, Japan, France, Slovenia, Spain, Italy, Greece, the Czech Republic, Estonia, Cyprus, Slovakia, Poland,

Lithuania, Argentina, Chile, Portugal, Hungary, Latvia and Croatia) (Figure 3.12). This group consists of countries with very high human development, which are fully free, and therefore in objective terms they clearly stand out in a positive sense among the DDCs. They represent something of an anomaly. What they have common is the conviction of their citizens, based upon diverse factors, that the living conditions in their countries are not (yet?) good enough. Some of this group are *rich and free LDCs* – the Republic of Korea, Italy, Greece, Slovakia, Lithuania, Hungary, Latvia and Croatia (Figure 3.13). In their case, a high quality of life is combined with high dissatisfaction with life, which may be symptomatic of a deep social crisis. On the other hand, the combination of a high quality of life with low to moderate social discontent (the rest of the rich and free DDCs) may be conducive to further dynamic development. Perhaps it is in this category, rich and free DDCs, which are not LDCs – comprising thirteen countries (the United States, Israel, Japan, France, Slovenia, Spain, the Czech Republic, Estonia, Cyprus, Poland, Argentina, Chile and Portugal) – that the developing countries of our day should also be sought (*rich and free developing countries*; see Figure 3.12).

Conclusions

One of Poland's most celebrated nineteenth-century painters created a monumental panorama of one of the greatest battles of medieval Europe (*The Battle of Grunwald* by Jan Matejko, 1875–1878). Bodies, scenes and symbols are strewn across a canvas spanning more than 40 m^2. Visitors to Warsaw's National Museum, who gaze up at the immense work of art, are initially overwhelmed by the complex and elaborate depiction. But then, in order to grasp the vivid mass of detail, they begin to pick out individual figures, episodes and symbols, and allow their minds to gradually make sense of the reality frozen on the canvas. In a similar fashion, the international community presents itself to our eyes full of overwhelming complexity and vivid with colours, shades and shapes, and each of us has the right to attempt to grasp for ourselves this tangle of events and relationships which is so difficult to understand and describe. In this case, however, unlike on the wall of the Warsaw museum, the scene is not frozen but continues to move dynamically, an ever-changing shape and internal structure. Hence the constant need for us to reject old concepts and frameworks, and construct new ones, going beyond accepted canons in the search for principles and categories by which to better comprehend the world.

Notes

1 Walt Rostow's nomenclature for the stages of economic development (Rostow 1960), provided the inspiration for the phase designations used here.

2 For example, answers to the questions concerning trust in the national government and confidence in the judicial system were not included in the Human Development Report 2015 profiles of twenty countries: Qatar, Saudi Arabia, the United Arab Emirates, Bahrain, Kuwait, Oman, Iran, Jordan, Algeria, China, Libya, Turkmenistan, Uzbekistan, Tajikistan, Syria, Laos, Cambodia, Myanmar, Rwanda and Burundi (HDR 2015). In 2016, according to the Freedom in the World index, all the aforementioned countries

have the status of 'Not Free' with the exception of Kuwait (which is 'Partly Free') (Freedom in the World 2016). Perhaps in some cases where data is lacking, there are possible alternative or supplementary explanations for this, other than political motives.

References

Brandt, W. *et al.* (Independent Commission on International Development Issues) (1980) *North–South: A Programme for Survival: The Report of the Independent Commission on International Development Issues under the Chairmanship of Willy Brandt.* London: Pan Books.
Boniface, P. (ed.) (2003) *Atlas des relations internationales.* Paris: Hatier.
Boyd, A. and Comenetz, J. (2007) *An Atlas of World Affairs,* 11th edn. London: Routledge.
Courtois, S., Werth, N., Panné, J.-L., Paczkowski, A., Bartošek, K. and Margolin, J.-L. (1999) *The Black Book of Communism: Crimes, Terror, Repression.* Mark Kramer (ed.), trans. Jonathan Murphy. Cambridge, MA: Harvard University Press.
Crosby, A.W. (2015) *Ecological Imperialism: The Biological Expansion of Europe, 900–1900,* 2nd edn. Cambridge: Cambridge University Press.
Desperak, J. and Balon, J. (2003) *Tablice geograficzne.* Warsaw: Świat Książki.
Domański, R. (2004) *Geografia ekonomiczna: Ujęcie dynamiczne.* Warsaw: Wydawnictwo Naukowe PWN.
Freedom in the World 2016. Anxious Dictators, Wavering Democracies: Global Freedom under Pressure. Freedom House. Retrieved 22 September 2017 from https://freedomhouse.org/report/freedom-world/freedom-world-2016.
HDR (Human Development Report) (2015). Work for Human Development. New York: United Nations Development Programme. Retrieved 22 September 2017 from http://hdr.undp.org/sites/default/files/2015_human_development_report.pdf.
Knox, P. and Agnew, J. (1990) *The Geography of the World Economy.* London: Edward Arnold.
Lacoste, Y. (1976) *Géographie du sous-développement: Géopolitique d'une crise.* 3rd edn. Paris: Presses universitaires de France.
Landes, D.S. (1998) *The Wealth and Poverty of Nations.* New York: WW Norton & Co.
Moyes, A. and Hayter, T. (eds) (1965) *World III: A Handbook on Developing Countries.* Oxford: Pergamon Press, The Macmillan Company.
Nouschi, M. (2003) *Petit atlas historique du 20e siècle.* Paris: Armand Colin.
Pawłowski, S. (1938) O renesansie geografii politycznej, in S. Pawłowski, *Geografia jako nauka i przedmiot nauczania.* Lwów: Książnica-Atlas.
Porębski, C. (2004) Umowa społeczna, in B. Szlachta (ed.), *Słownik społeczny.* Kraków: Wydawnictwo WAM.
Rostow, W.W. (1960) *The Stages of Economic Growth: A Non-Communist Manifesto.* Cambridge: Cambridge University Press.
Sauvy, A. (1952) Trois mondes, une planète. *L'Observateur* 118: 5.
Sauvy, A. (1961) Le 'Tiers-Monde': Sous-développement et développement. (Réédition augmentée d'une mise à jour par Alfred Sauvy). Travaux et Documents. Cahier no 39. Paris: Institut national d'études démographiques, Presses Universitaires de France.
Solarz, M.W. (2012) 'Third World': The 60th Anniversary of a Concept that Changed History. *Third World Quarterly* 33(9): 1561–1573.
Solarz, M.W. (2014) *The Language of Global Development: A Misleading Geography.* Abingdon: Routledge.
Solarz, M.W. (2017) The Birth and Development of the Language of Global Development in Light of Trends in Global Population, International Politics, Economics and Globalization. *Third World Quarterly* 38(8): 1753–1766.
Swift, J. (2003) *The Palgrave Concise Historical Atlas of the Cold War.* Basingstoke: Palgrave Macmillan.
Wolf-Phillips, L. (1979) Why Third World?. *Third World Quarterly* 1(1): 105–115.

4 Twenty-first-century cities

From global challenges to local responses

Voicu Bodocan, Jozsef Benedek and Raularian Rusu

This chapter deals with globalized urban space and focuses on processes of post-socialist urban transformation in Central and Eastern Europe (CEE), now part of the European Union, and is organized in two parts. The first presents a general geographical perspective on urban processes and transformations on a global scale. We contend that both world cities and capital cities have played a key role in the post-socialist transformation in CEE. This part concludes with the presentation of a bottom-up approach for evaluating transformation in urban spaces, highlighting major trends in urban policy. The second part of the chapter addresses the special case of CEE cities and follows the same thematic sequence as the first: the role of capital cities in the post-socialist transition, the economic and social transformation of cities with a focus on suburbanization and urban sprawl, and finally the particular features of urban policy in CEE. Our purpose is to offer a regional synthesis rather than generate new empirical findings.

Global challenges: the perspective of urban geography

World cities and capital cities

The liberalization of the world economy has reconfigured global flows towards new sites of polarization known as *world* or *global cities*. Although there is a huge body of literature revealing the dominance of non-global processes in urban growth (Jacobs 2013), the world city theory prevails in the international literature of the *new urban geography*. The work of Friedmann, Sassen and Taylor, as the main architects of this theory, was developed later within the influential Loughborough University based Globalization and World Cities Research Network (GaWC). Most definitions consider global cities as '*command and control centres*' of the global economy (Sassen 1991; Friedmann 1995; Pacione 2009; Taylor and Csomós 2012). They are not identified by size but by their function and interaction, and their coordination of the global economy. They form a distinctive class of cities (Taylor 2000) with many common features that differentiate them from others in how they look, function and socially interact. Their main characteristics derive from their cross-border relations (Parnreiter 2014), cosmopolitan culture (Friedmann 1995), transnational professional class (Sassen 2007), segregated neighbourhoods and 'monumental skylines' (Ford 1998).

World city formation is related to deindustrialization and the new spatial division of labour, which have led to an accumulation of capital in the new cores of the global system. These cores control important regional or national economies and articulate them with the global economy (Friedmann 1995). They also 'transmit the impulses of globalization back to national and provincial centers' (Knox 2014: 13). It is an ongoing process, amplified by the increasing networking of cities and firms (Liu, Derudder, Witlox and Hoyler. 2014a).

Due to the complexity of their operations, multinational companies have outsourced a substantial number of their services to specialized firms, themselves also multinationals, which follow them around the world (Derudder and Witlox 2008). This geographical dispersion has resulted in a geographical segregation of control determined by the overconcentration of headquarters in just a few cities. The office networks developed by the specialized multinationals, and the capital, people and information flows through them are responsible for world city formation because they interlock cities (Derudder and Parnreiter 2014) and foster their mutual cooperation by communication, transfer of knowledge, transactions and face-to-face meetings. These firms are identified by Sassen (1991) as providers of *advanced producer services*, and grouped by Taylor (2000) into four complementary sectors: finance and banking, accounting, advertising and law. Such companies 'map' world cities much as airlines determine the existence and importance of their hubs. This connectivity analysis has also been extended to other actors, which can produce international flows (cultural, academic, tourist etc). World cities cannot be viewed as individual entities, but only in relation to other cities within an *interlocking network* (Taylor 2001) connecting the world economy, companies and cities.

These international flows have been supported by the development of *transportation*, which has followed a parallel path to that of world city formation. Keeling (1995) was one of the first to consider air links in assessing the interaction of world cities. Airports are not only hubs facilitating the international flow of passengers, or locational determinants for new investments, but they are also prerequisites for development based on the knowledge and information economy. The offices of advanced producer service firms are likely to be located in those cities with many air transport links (Liu, Derudder and Gago García 2013), and they have their headquarters in 'the world's best connected places' (Bassens and Van Meeteren 2015: 756).

Although international flows have been crucial in the formation of world cities, *locational attributes* are equally important. Market centrality, intermediacy and accessibility, all reflect the power of location, and the primary world cities record high values in terms of these attributes. These cities represent unique knowledge complexes (Knox 1995), where the local cultural environment has had an imprint on the professional experience and knowledge of the multinationals. Despite the fact that capital accumulation has taken place in a limited number of spaces, Friedmann (1995) and Taylor (2004) consider that all cities are able to connect to the network. Most world cities belong to federal and decentralized states with deregulated economies, as opposed to many unitary states where political control prevails. Global cities are large urban agglomerations, with significant growth in the previous phases of globalization and belong mostly to advanced or large economies (e.g. the BRIC economies).

Besides the commercial network as systemic in the contemporary world system, Taylor (2000) also identifies a *diplomatic network* developed by inter-state world city relations, where capital cities are major actors as 'political command and control centres'. The interaction between nation states, cities and the global economy, in terms of power and control, have become an important subject in political geography and the role of capital cities can reveal potential explanations in *city/state relations* which have recently changed.

In the traditional approach, *capital cities* are related to political, economic, cultural and demographic centrality, and have historically belonged to the world's most important cities because they are centres of state power in a context where political control has been dominant in world affairs. They played a crucial role in the foundation of the nation state, and they possess different types of political and economical control. However, the authority of the state and the primacy of capitals have come under threat from contemporary processes at three levels: decentralization policies at a local level, supranational organizations (like EU or IMF) at a regional level and the rising importance of world cities at a global level. Indeed, world cities tend to avoid national connections (Abrahamson 2004), and the state is losing control both over its internal and external affairs. The territorial state is being threatened by transnational power (Friedmann and Wolff 1982), which is taking away its functions (Taylor 1993), therefore, the geography of political power could turn its focus to the ascending global cities as alternatives to capital cities (Campbell 2000).

When the economic network intersected the diplomatic network, it gave rise to the world's most powerful cities, namely, national capitals as world cities. Their primacy may have been accentuated by the globalization process (Taylor and Derudder 2014). Despite the changing role of capital cities, twenty-eight out of forty-eight 'Alpha' world cities as ranked by GaWC in 2012, have or have had this status of political control at national or provincial level. Capitals, through their institutions, rule territories and people and 'possess sui generis qualities, including unique political, cultural and administrative functions' (Drewe 1993: 368). They are places where government and private sectors interact (Mayer, Sager, Kaufmann and Warland 2016) in a global context in which private domestic investment is four times larger than foreign direct investment (McCann and Acs 2011).

Transformations in urban space: suburbanization, urban sprawl and social change

Increasing urbanization has forced cities to develop both horizontally and vertically. One of the most striking forms of horizontal urban expansion is represented by *suburbanization*, which involves an urban structure with a central city core, inner suburbs and outer suburbs (Zhou and Ma 2000). The definition of 'suburbs' varies, as they may be based on administrative criteria (the inner suburbs are within the administrative limits of the city, while the outer suburbs are outside these limits) or functional ones, where the inner suburbs consist mainly of residential, commercial and industrial buildings, while the outer suburbs are especially residential areas, whose inhabitants commute to the city and its inner suburbs, or are engaged in agricultural activities.

The suburbs are no longer necessarily inferior to the city core. They have increasing population numbers, economic power and employment opportunities, and have become important industrial, commercial and administrative centres, sometimes effectively competing with the central city. New spatial structures emerge as central cities make way for polycentric urban areas consisting of outer cities, suburban cities and edge cities (Bruegmann 2005).

The driving forces of suburbanization vary around the world. However, at their root one may generally perceive economic factors (lower costs of living, land use and construction in the suburbs) along with social and environmental concerns (the desire for increased living space, less pollution, better recreational and outdoor facilities, higher security). The transport issues associated with suburban life have been largely overcome thanks to the widespread use of private automobiles, the continuous expansion and development of motorways and highways, as well as better and faster public transport available to commuters (Anas and Pines 2008).

Excessive or chaotic growth of suburban areas may lead to *urban sprawl*. Criticism of urban sprawl is based on the assumption that open space has an aesthetic and environmental value. Also urban sprawl diminishes an allegedly scarce resource, namely, farmland. Although in the Western world there is free market competition between urban land use and agricultural use, usually the outcome is in favour of the former. Excessive commuting may cause traffic congestion and higher pollution, in response to which the building of new infrastructure may be deemed necessary (Brueckner 2000). Other negative consequences are the degradation of central-city downtowns (as investments are directed to the suburbs), as well as the social alienation and isolation of suburbanites who do not form socially integrated communities.

The historic picture in the developed world was that those moving out from the city centres were mainly members of the middle and upper classes, who could afford cars and wanted better housing and living standards for their families. Population growth in the cities was less important as a factor, as sprawl emerged even in declining urban areas (Herzog 2014).

By contrast, in the developing countries, sprawl has been related mainly to population growth in the cities, basically due to migration from the countryside and the inability of migrants to find housing at accessible prices in the city centre. Therefore, in these contexts, sprawl has taken the form of shanty towns (Herzog 2014). As a consequence, urban sprawl also produces social and even racial segregation (Hamel and Keil 2015).

However, there is also a process of *reurbanization*, as described by Rérat (2012), which means the growth of the central city irrespective of the growth of its suburbs. Population increase in the central city is mainly due to international migration and changes which occur in the lifestyles of individuals, as young (especially unmarried or childless) adults and the elderly prefer a city centre environment to the suburbs, unlike families with children, who often move to the suburbs in search for more space and a quieter environment (Rérat 2012). In Bruegmann's opinion (2005), *gentrification* and urban sprawl are just the flip sides of the same coin.

Recent trends in urban policy and urbanization

In an increasingly urbanized world, cities are seeking different urban manage-ment methods and urban policies in line with their resources, traditions and the local and national contexts in which they are embedded. Globalization and inten-sified inter-urban competition for the attraction of scarce development resources like capital, investment and labour have led to a variety of adaptation strategies. In other words, cities develop and position themselves in the global flow of capital through the formulation of a wide range of strategies and policies which affect the welfare of residents and the characteristics of the urban area.

Generally, there are two ways to design local responses to the changing eco-nomic, social and political environment. The first is represented by *urban policy* (locational policies, urban planning) developed by local government authorities, and the second is represented by *national policies designed for urban areas*.

In this policy framework, cities are considered as important drivers of economic growth; they are the solutions to social problems rather than the cause of them. A special role is attributed to capital cities as promoters of unique regional innova-tion systems (RIS) through locational policies, which position them in the national urban hierarchies (Mayer *et al.* 2016) and increase their international competit-iveness. In this context, innovation is considered as an important source of com-petitive advantage for cities. The RIS literature on policy making presents a conservative view on policy adoption, stating that public policies should be developed only to address market and system failure (Asheim, Boschma and Cooke 2011). Recently, Mayer *et al.* (2016) have argued for an alternative, broader concept of locational policies, encompassing four distinct categories: innovation policies, coordination, attracting and soliciting ('asking for') money. Based on certain location-specific advantages and assets (skilled labour, accessibility, quality of environment, etc), their aim is to attract capital investment to cities and enhance their economic competitiveness. Therefore, the interplay between the economic sector structure of cities and the local institutional setting will determine the policy outcome for each locality. More specifically, innovation policies encourage inter-actions between different local actors engaged in knowledge production: firms, universities, local government, in order to create knowledge spillovers.

Strategies for attracting money target the enlargement of local tax revenues by increasing the number of firms and taxpayers. Asking for money focuses on the preservative role of certain special functions – carried out by, for example, capital cities or university towns – to justify redistribution or compensation payments. Top-tier cities (capitals, regional centres) have important entrepreneurial potential, which can be developed within a RIS and thus can implement strong money attraction strategies, while second-tier cities can emphasize more the specific role they play. Location policies can be inward- or outward-oriented; in the latter case, they are positioning strategies (Mayer at al. 2016).

The changing economy of urban agglomerations, the decline of large-scale manufacturing and emergence of new service industries is putting constant pressure on local government to scale up development strategies and policies. The increasing mobility of workers and residents has led to a reconsideration of the role of the residential economy, which has become a strategic factor in the generation of income within regions. Contrary to export-based theories, regions with a higher income are those with a strong residential economy (Mayer at al. 2016).

Local responses to global challenges: the case of CEE

Capitals and world cities in post-socialist Europe

Anti-globalization policies had limited urban growth before 1990 and the networking of cities was almost non-existent, within CEE or between the disjointed urban systems of Europe. Therefore, cities in the post-socialist CEE states had to pass through a quick transition, from a planned and centralized system to a new economy based on the globalization of production and services. Their connection to international flows of capital and knowledge was made possible by attracting foreign funds and investments in order to exploit the advantages of CEE cities in terms of human capital quality and labour costs. These were important locational determinants for multinationals seeking to expand their network of production sites and service offices in the CEE area. The opening of new markets and privatization of state-owned companies was an opportunity for advanced producer services to deploy in the region. In this way, the CEE cities became space for both core and peripheral processes due to the de-industrialization of Western manufacturing economies and the delocalization of production and services by multinationals.

Specific to this region was the reconfiguration of the *territorial control* of certain capitals because they had smaller economies to articulate (Prague) or new ones to command (Ljubljana, Bratislava, Riga, Tallinn, Vilnius and Zagreb). Concurrently, new actors were involved in territorial decision making as decentralization policies were adopted requiring the new local authorities to engage in attracting investments and promoting their cities (Robinson 2005).

Competition for investment has resulted in a polarization and segregation that has accentuated existing *discrepancies* such as territorial inequalities related to the historical East–West divide (e.g. Poland, Romania) and increased the primacy of the capitals within inter-city rivalry. Foreign investments were directed preferentially in the 1990s towards capital cities, which benefitted the most by far from new accumulation. The capital cities of CEE became gateways (Csomós 2011) to which the multinationals started to transfer their regional headquarters during their expansion into the newly emerging economies.

The evolution of CEE's *regional connectivity* to the world city network has been characterized by an 'uneven spatial pattern of globalization' (Hamilton and Carter 2005: 148) because of the divergent opportunities which the region's cities had to connect to the world system due to historical factors, geographical location and cultural interaction. Unlike established world cities, analysis of the potential of world city formation in the region should be considered in relation to the limited time available and the different pace of integration the post-socialist cities have followed, depending on the level of economic restructuring and political decentralization. At the same time, rapid growth of connectivity with global flows has been facilitated by integration into the European Union (Hamilton 2005), an asset of these cities compared to those in other formerly centralized state economies.

There are *fifteen world cities* located in CEE according to the 2012 GaWC hierarchy, fourteen of them being capitals and one a historic capital (Krakow). This reflects the significant roles of these cities. Warsaw has been able to surpass traditional financial centres (Clark 2015) and become the highest ranked world city in CEE. While Warsaw's position reflects the dimension of its market and economy, Prague is the best located capital, and Budapest has taken advantage of its centrality within an expanding market. Transport infrastructure is a key element in this hierarchy of cities, favouring those with higher values in air traffic and motorway connectivity. Most of the new capitals (Zagreb, Ljubljana, Bratislava, Vilnius, Riga and Tallinn) are small world cities but their demo-graphic and economic lack of size is compensated for by performance and Euro-pean Union integration. Capitals like Bucharest and Sofia, secondary world cities with a peripheral and non-contiguous position in relation to the Western states, have recently made up a good part of the gap (Liu, Derudder and Taylor 2014b). They have become more advanced than their state economies and much closer to the capitals of Central Europe (Bourdeau-Lepage 2007).

As states remain here very powerful and continue to 'shape the market' (Taylor and Derudder 2016: 73), capital cities in CEE have strategic positions as *centres of decision making* and interfaces between the national and global economy, responsible for the integration of their regions into global capital. Except for Poland, the CEE states have not advanced a second tier of world cities, which indicates the domination in those countries of a centralized economy and monocentric structure of development with a primate city. In many of these countries, secondary cities have mainly remained destinations for manu-facturing investment as part of reindustrialization (Sýkora and Bouzarovski 2012), or spaces for developing creative industries (Becuţ 2016). These cities can evolve faster and connect more easily to global flows if they are accessible and have local political and business elites able to attract knowledge resources and educated people and thus develop a creative and cultural economy.

Urban transformation in CEE

While the political and economic transition may be over, post-socialist cities are still in transition (Sýkora and Bouzarovski 2012). In the 1990s, urban transformation was generally left in the hands of the free – that is, unregulated – market, in line with prevailing neoliberal ideology. Cities experienced dramatic *transformations* as the central state power was no longer involved in their development and the market economy became the driving force behind their expansion (Hamilton, Dimitrovska Andrews and Pichler-Milanović 2005).

During the socialist period, while most Western cities began to deconcentrate, urbanization in the cities of the CEE countries was accelerated by the building of high-density socialist housing units, functionally integrated with industrial zones and compact built-up urban areas for reasons of economic efficiency and social control (Hirt 2013; Sýkora and Stanilov 2014). Suburban development was minor and heavily restricted, as in the case, for instance, of *dacha* housing areas.

After 1989, the restoration of land to private owners triggered the emergence of new forms and patterns of real estate development (Smith and Timár 2010). In the Czech Republic and Estonia, properties were rather quickly returned to their pre-socialist owners, providing a significant supply of real estate for the property market, followed (at least in Prague and Tallinn) by commercialization and gentrification. In contrast, this path was not followed by Budapest and other cities in Hungary and Romania (Enyedi and Kovács 2006; Sýkora and Bouzarovski 2012).

Furthermore, the arrival of international capital in the emerging markets of the CEE countries created the need for new residential, retail, industrial and office premises, concentrated initially in the main urban areas and increasingly also in the suburbs (Sýkora and Stanilov 2014). Evidence shows that, at least since the mid-1990s, *sprawl-type suburbanization* has become the main form of urban growth in the post-socialist cities of CEE. The middle and upper classes emerging in the last twenty-five years demand higher standards of housing, and this has led to the movement of an ever-increasing share of the urban population to suburban family homes dispersed around the compact city cores (Hirt 2013).

Post-socialist suburbanization is also promoted by local authorities, eager to create opportunities for economic development and any sort of growth. Therefore, the decentralization of power is the main cause of urban sprawl characterized by chaotic development and fragmented spatial patterns (Sýkora and Stanilov 2014). Urban hinterlands are radically transformed through commercial and residential suburbanization.

At the same time, the *inner cities* are confronted with the physical decay of housing stock, and sometimes pollution due to still active industrial units as well as archaic heating systems. While the younger and higher-educated generations move out to the suburbs, city cores become more and more dominated by an ageing population (Enyedi and Kovács 2006). On the other hand, the commercialization and expansion of city cores, as well as the dynamic revitalization of some neighbourhoods lead to visible restructuring effects and an attractive urban landscape in certain cities (Sýkora and Bouzarovski 2012). Urban restructuring also involves *gentrification*, brownfield regeneration, the building of new residential areas, office clusters and shopping centres. In the case of some Polish and Czech cities (e.g. Łódź and Brno), there has been a noticeable rejuvenation of the city centres, whose central location, cultural facilities and urban lifestyle are magnets for 'transitory urbanites', mainly university educated young people. In a way, this group is similar to the gentrifiers in the Western world, contributing to a 'symbolic upgrading and an increasing appreciation of inner-city housing and living' (Haase, Grossmann and Steinführer 2012: 324). One negative side has been the shrinkage of green areas in many inner cities, for example Sofia, as common space has been converted to private use (Hirt 2013). The surviving green belts around the large cities have been even more impacted by the building of neighbourhoods for the nouveau riche (e.g. in Bucharest) (see Box 4.1).

Cities' response to change: urban resilience

The development of most cities is influenced by urban policy and planning, and this correlation was extremely strong in CEE during the socialist period of urbanization and industrialization, when state-led investment and regulations created a distinctive pattern of land use, with high-density new residential areas. Of course, after 1989, urban policy and planning underwent a process of Europeanization and democratization, in which its role was diminished in comparison with the socialist period, but many of the urban conflicts and problems currently experienced in CEE are inherited from that time of transition. The cities of CEE responded variously to the urban change induced by the system transformation from socialism to capitalism, according to their national context, internal adaptation capacities, such as skills, infrastructure or natural assets, and the availability of new windows of opportunity resulting from their integration into international flows of capital, technology and labour. In addition, accession to the European Union created new opportunities to attract funds for urban development.

Local governments as actors of urban development. The importance and role of urban policy in CEE largely depends on the institutional arrangement appropriate to the administrative-territorial organization of each country and the division of functional responsibilities between local and central governments. All the CEE countries are unitary states, therefore there is little difference across the countries in terms of the importance of local government. This means that local governments play a more important role in service provision compared to federal states (Wolman and McManmon 2012), where there is a strong intermediate regional level of government fulfilling this role. Poland has a particular situation in this respect due to the stronger role of its self-elected regional governments in promoting and implementing services and economic development. According to Wolman and McManmon (2012), in most CEE countries, local government provides the majority of funding for services like education, environmental protection, housing and community amenities (more than 50 per cent of the local expenditure of all government spending on each service in 2007), while the

Box 4.1 Suburbanization and urban sprawl in Romania. Case study: Florești (Cluj-Napoca metropolitan area)

Suburbanization is not necessarily a recent process in Romania but it reached new heights after the end of the communist regime. There was some sort of suburbanization even before 1990: many cities were surrounded by areas officially designated as suburban communes, which, however, lost this status in the aftermath of the events of December 1989. Moreover, after decades of encouraging emigration from rural areas to cities, in the 1980s, the communist authorities decided to block this outflow because of the lack of workforce in agriculture and the steep decline in the rural population. Rural migrants were prevented from moving to cities except in specific cases, although one possibility was migration to the suburban communes, where commuting was an easy option (Hirt 2013). As a result, many of these areas recorded some population growth even during the 1980s.

The restitution of private land to the original owners or their heirs, as well as the transformation of the political system and the introduction of the market economy changed Romanians' perspectives and opportunities in terms of real estate ownership. While the newly gained 'freedom' allowed many people to buy the much-desired flat in the city, aspirations soon changed, especially for young, middle-class, upper-class and university educated city dwellers. Especially after 2000, when the Romanian economy slowly recovered after the most difficult years of transition, many people in the cities, especially young families with children, began to search for a home in a quieter environment. Modern houses emerged in the countryside, mainly in villages with good road connections to the city, and car ownership became the necessary condition for such a move. While this happened on a small scale almost everywhere in Romania, the massive construction of new housing areas is characteristic for the commune of Florești, only 8 km from Cluj-Napoca city centre, which had 5,868 inhabitants in 1990 and a slightly increased figure of 7,018 in 2000, but by 2014 it had reached 23,632 inhabitants, which makes it the largest commune in Romania in terms of population.

What are the reasons behind such a housing development? Certainly, the village of Florești was already attractive to Cluj-Napoca city dwellers in the 1990s and the beginning of the 2000s, but its rather slow growth was typical of other villages surrounding the city. This can be easily demonstrated by the positive migration balance that characterized the commune from 1992 onwards. Migration values began to rise steeply in the 2000s, initially because of the development of a new neighbourhood built by the local authorities. This was an estate of 'ANL' (i.e. state-supported) blocks of flats intended as social housing for young people, especially young families with children, who could not afford to live in the city of Cluj-Napoca. Low rents and the opportunity to buy the flat after a certain period at a low price led to very high demand.

Real estate developers soon realized that there was ample cheap land available between the new housing estate and the village centre (on the main road). With the help of the local authorities, other housing areas emerged in the area. The private developers provided competitive prices, significantly cheaper than those in Cluj-Napoca, attracting thousands of new inhabitants to Florești. The migration balance grew from 36.04 per thousand in 2006 to 73.45 per thousand in 2007, and to 160.82 per thousand in 2008, reaching a maximum of 270.41 per thousand in 2010.

It has remained between 100 and 200 per thousand annually since then (INS 1998–2016).

Apart from the original blocks of flats, other types of housing appeared, such as individual and multi-family houses. This involved a diversification of supply, which meant that the neighbourhoods developed chaotically, a feature which was strongly criticized by urban planners and other experts, both on aesthetic and technical grounds. The newly developed housing estates led to a higher population density than had been estimated, resulting in gas and power supply difficulties, as well as strain on the water and sewerage systems. Moreover, most inhabitants are commuters and there is a lack of road infrastructure. The only road to Cluj-Napoca is inadequate and alternative means of transport are unreliable, as public transport is based only on buses and these are used only by a minority of residents. The result is severe rush-hour traffic congestion.

While urban sprawl is not typically associated with high-density housing, the rather unregulated real estate development in Florești may be characterized as sprawl due to the degree of its open space consumption, the lack of any pattern of development and the inability of the local authorities to tackle the associated infrastructure problems.

Czech Republic and Poland have an unusually high local funding of recreation, culture and religion (more than 70 per cent). However, it is worth noting the low share of local spending on economic affairs and social protection (less than 50 per cent) in all the countries, which means that local governments are less able than national governments to design policies aimed at increasing income and enhancing the well-being of local residents, with one notable exception: Poland (47.6 per cent).

Urban planning. During the post-socialist transition period, local economic development became dominated by private capital, therefore urban policies related to economic development went to indirect instruments of urban planning in order to create an attractive urban environment for private capital. This turn was determined by the rapid privatization of state-owned firms and by the change in the property structure of economic enterprises. The primary objective of urban planning in CEE is to regulate land use and development in urban areas. However, in the early transition period, these countries adopted a neoliberal position towards planning, reducing to a minimum government intervention in land development, and allowing the establishment of a weak planning environment. The state planning control of zoning regulations together with real-estate taxes are tools of local government intended to influence urban development in the general public interest. Under socialism, the housing market was limited to its 'shelter' functions, while its 'investment' function played a minor role in the majority of CEE countries. Post-socialist privatization led to the opposite situation in some countries, like Romania and Hungary, where urban land and housing have been completely privatized. Generally, residential property in the big cities is costly, reflecting the capitalization of the urban infrastructure in housing prices. In parallel, local authorities are facing the multiple challenges of post-industrial restructuring, privatization of the service sector, suburbanization, increased residential mobility, the opening up of cities to foreign investment, increasing social polarization and international migration.

National policies towards cities. The strong involvement of central government in the design of urban policies is specific to the CEE countries. Two policies have been followed during recent decades in many of these countries: the creation of new cities by urban reclassification and support for urban growth poles. In a context characterized by a general decline in the population as a whole (between 1991 and 2015), the urbanization rate of the CEE countries has shown a general trend towards an increasing share of the urban population in the total (Table 4.1). In some cases (e.g. Hungary and Romania), urbanization has been viewed as a spatial strategy promoting growth and well-being through the expansion of urban services. Therefore, a large campaign of urban reclassification has been carried out (see Box 4.2).

As a result of the increasing Europeanization of spatial planning, CEE countries have taken over the principles and major objectives of the relevant European spatial development documents. Among these objectives, the development of a balanced and polycentric spatial structure ranks highly and this has become a key priority in implementing spatial policy objectives for major transportation infrastructure development and also urban development. One important policy aimed at balancing spatial development has focused on encouraging growth poles (Table 4.1) (see Box 4.3), which were devised, implemented and contested in France some fifty years ago, but have now rediscovered and introduced into European Spatial Development Policy.

Box 4.2 Post-socialist urban policy: new cities in Hungary

Hungarian settlement distribution is highly polarized by the capital city Budapest, which accounts for around 20 per cent of the country's population. New, post-socialist urbanization has been promoted by the central government as a regional development policy, and has led to the second highest level of urbanization in CEE (71 per cent, versus the Czech Republic's 73 per cent), but this is viewed by many as a merely statistical exercise. The process follows certain benchmarks determined by the central government, which have to be fulfilled by rural settlements in order to receive urban status. However, as in Romania, the new urbanization is fuelled not only by developmental considerations, but also political power games, to the benefit of local and national political actors. It is facilitated by the existence of an important pro-urban and anti-rural legacy, which considers urban places and life as superior to rural existence (Kulcsár and Brown 2011). After the change of regime, the number of towns and cities increased from 166 in 1990 to 346 in 2013. However, the new urban settlements were not able to improve their demographic, social and economic performance. Under these circumstances, a crucial question is: how can urbanization strategy be shaped in a way that really contributes to the development of reclassified settlements?

Table 4.1 Total and urban population in CEE

	Total population 1991 ('000)	Total population 2015 ('000)	Urban population 1991 ('000)	Urban population 2015 ('000)	Urbanization rate 2015 (%)
Czech Republic	10,330	10,777	7,764	7,866	73
Hungary	10,370	9,911	6,815	7,060	71
Poland	38,255	38,222	23,456	23,139	61
Romania	23,356	21,579	12,596	11,774	55
Slovakia	5,298	5,458	3,008	2,925	54
Estonia	1,549	1,280	1,101	865	68
Latvia	2,645	2,031	1,830	1,369	67
Lithuania	3,699	2,999	2,498	1,995	67
Croatia	4,799	4,255	2,608	2,509	59
Slovenia	2,005	2,079	1,013	1,032	50
Total	102,306	98,591	62,689	60,534	61

Source: own compilation based on United Nations statistics (2015).

Box 4.3 Promotion of urban growth poles: a Romanian case study

Romania is the second largest country in CEE, with significant regional disparities. Driven by the growth of the capital region Bucharest-Ilfov, regional polarization is at a historical high and Romania has become the most spatially unequal country in CEE. The capital region accounts for 10 per cent of the national population, but more than 50 per cent of foreign direct investment, and around 25 per cent of national GDP. In order to counterbalance the strong spatial polarization of development, Romanian spatial planning policy seeks to channel growth towards a larger number of regional centres, and by so doing enhance the polycentric spatial structure of the country. Therefore, during the last programming period, an important part of regional policy concentrated on delineating and supporting seven growth poles. Accordingly, the Regional Operational Programme (ROP) 2007–2013 set out five specific aims: increasing the social and economic importance of cities (through polycentric development); providing better access to regions (by improving public transportation in cities and their surrounding areas); improving regional social infrastructure; enhancing regional competitiveness; and increasing the regional economic importance of tourism. It is noteworthy that all except one of these (tourism) targeted cities. Budget allocation according to the stated framework created an urban concentration of funding resources, which actually contributed to an increase in regional disparities. As a consequence, although Romania fully met its most important territorial development goal – EU convergence (the GDP per capita compared to the EU average increased from 26 per cent in 2000 to 49 per cent in 2012) – it did so at the cost of increased internal territorial polarization. The designation of urban growth poles and the concentration of ROP resources in cities have significantly contributed to this situation, which exemplifies the problem of development policy instruments being taken over and applied without regard to the national and spatial context and specificities.

Conclusions

We have tried to demonstrate that the CEE countries have experienced a particular urban transformation, marked by a rapid system change from socialism to capitalism. The capital cities have been the leading force of the urban transformation, as they have accounted for a significant amount of national economic growth. On the one hand, the strong spatial concentration of development has created unprecedented regional inequalities, while on the other hand, the capital cities themselves have experienced strong internal differentiation and social segregation. A specific form of sprawl-type suburbanization has taken place in CEE, while urban policies are still strongly dominated by central government led national policies, excepting Poland, which has strong regional and local administration.

References

Abrahamson, M. (2004) *Global Cities*. New York: Oxford University Press.

Anas, A. and Pines, D. (2008) Anti-Sprawl Policies in a System of Congested Cities. *Regional Science and Urban Economics* 38: 408–423.

Asheim, B.T., Boschma, R. and Cooke, P. (2011) Constructing Regional Advantage: Platform Policies Based on Related Variety and Differentiated Knowledge Bases. *Regional Studies* 45(7): 893–904.

Bassens, D. and Van Meeteren, M. (2015) World Cities Under Conditions of Financialized Globalization: Towards an Augmented World City Hypothesis. *Progress in Human Geography* 39(6): 752–775.

Becuţ, A.G. (2016) Dynamics of Creative Industries in a Post-Communist Society: The Development of Creative Sector in Romanian Cities. *City, Culture and Society* 7: 63–68.

Bourdeau-Lepage, L. (2007) Advanced Services and City Globalization on the Eastern Fringe of Europe. *Belgeo* 1: 133–146.

Brueckner, J.C. (2000) Urban Sprawl: Diagnosis and Remedies. *International Regional Science Review* 23(2): 160–171.

Bruegmann, R. (2005) *Sprawl: A Compact History*. Chicago, IL: University of Chicago Press.

Campbell, S. (2000) The Changing Role and Identity of Capital Cities in the Global Era. Paper presented at the Association of American Geographers Annual Meeting, Pittsburgh, PA, 4–8 April 2000.

Clark, G. (2015) *The Making of a World City: London 1991 to 2021*. Chichester: John Wiley & Sons.

Cochrane, A. (2016) Urban Economics and Urban Policy: Challenging Conventional Policy Wisdom. *Regional Studies* 50(8): 1465–1467.

Csomós, G. (2011) Analysis of Leading Cities in Central Europe: Control of Regional Economy. *Bulletin of Geography: Socio-Economic Series* 16: 21–33.

Derudder, B. and Parnreiter, Ch. (2014) Introduction: The Interlocking Network Model for Studying Urban Networks: Outline, Potential, Critiques, and Ways Forward. *Tijdschrift voor Economische en Sociale Geografie* 105(4): 373–386.

Derudder, B. and Witlox, F. (2008) Mapping World City Networks Through Airline Flows: Context, Relevance, and Problems Mapping. *Journal of Transport Geography* 16: 305–312.

Drewe, P. (1993) Capital Cities in Europe: Directions for the Future, in J. Taylor, J.G. Lengelee, Caroline Andrew (eds), *Capital Cities/Les Capitales: Perspectives Internationales/International Perspectives*. Ottawa: Carleton University Press, 343–377.

Enyedi, G. and Kovács, Z. (eds) (2006) *Social Changes and Social Sustainability in Historical Urban Centres: The Case of Central Europe*. Pécs: Centre for Regional Studies of the Hungarian Academy of Sciences.

Ford, L.R. (1998) Midtowns, Megastructures and World Cities. *The Geographical Review* 88(4): 528–547.

Friedman, J. and Wolff, G. (1982) World City Formation: An Agenda for Research and Action. *International Journal of Urban and Regional Research* 3: 309–344.

Friedmann, J. (1995) Where We Stand: A Decade of World City Research, in P.L. Knox and P.J. Taylor (eds), *World Cities in a World-System*. Cambridge: Cambridge University Press, 21–48.

Haase, A., Grossmann, K. and Steinführer, A. (2012) Transitory Urbanites: New Actors of Residential Change in Polish and Czech Inner Cities. *Cities* 29: 318–326.

Hamel, P. and Keil, R. (eds) (2015) *Suburban Governance: A Global View*. Toronto: University of Toronto Press.

Hamilton F.E.I. (2005) The External Forces: Towards Globalization and European Integration, in F.E.I. Hamilton, K. Dimitrovska Andrews and N. Pichler-Milanović (eds), *Transformation of Cities in Central and Eastern Europe: Towards Globalization*. Tokyo: United Nations University Press, 79–115.

Hamilton, F.E.I. and Carter, F.W. (2005) Foreign Directs Investments In City Restructuring, in F.E.I. Hamilton, K. Dimitrovska Andrews and N. Pichler-Milanović (eds), *Transformation of Cities in Central and Eastern Europe: Towards Globalization*. Tokyo: United Nations University Press, 116–152.

Hamilton, F.E.I., Dimitrovska Andrews, K. and Pichler-Milanović, N. (eds) (2005) *Transformation of Cities in Central and Eastern Europe: Towards Globalization*. Tokyo: United Nations University Press.

Herzog, L. (2014) *Global Suburbs: Urban Sprawl from the Rio Grande to Rio de Janeiro*. New York: Routledge.

Hirt, S. (2013) Whatever Happened to the (Post)Socialist City?. *Cities* 32: 529–538.

INS – Institutul National de Statistica, Romania 1998–2016. Retrieved 8 August 2016 from http://statistici.insse.ro/shop/index.jsp?page=tempo2&lang=ro&context=12.

Jacobs, A.J. (2013) Preface to *The World's Cities: Contrasting Regional, National and Global Perspectives*. New York: Routledge.

Keeling, D.J. (1995) Transport and World City Paradigm, in P.L. Knox and P.J. Taylor (eds), *World Cities in a World-System*. Cambridge: Cambridge University Press, 115–132.

Knox, P. (2014) Introduction, in P. Knox (ed.), *Atlas of Cities*. Princeton, NJ: Princeton University Press, 10–16.

Knox, P.L. (1995), World Cities in a World-System, in P.L. Knox and P.J. Taylor (eds), *World Cities in a World-System*. Cambridge: Cambridge University Press, 3–21.

Kulcsár, L.J. and Brown, D.L. (2011) The Political Economy of Urban Reclassification in Post-Socialist Hungary. *Regional Studies* 45(4): 479–490.

Liu, X., Derudder, B. and Gago García, C. (2013) Exploring the Co-Evolution of the Geographies of Air Transport Aviation and Corporate Networks. *Journal of Transport Geography* 30: 26–36.

Liu, X., Derudder, B., Witlox, F. and Hoyler, M. (2014a) Cities as Networks Within Networks of Cities: The Evolution of the City/Firm-Duality in the World City Network, 2000–2010. *Tijdschrift voor Economische en Sociale Geografie* 105(4): 465–482.

Liu, X., Derudder, B. and Taylor, P.J. (2014b) Mapping the Evolution of Hierarchical and Regional Tendencies in the World City Network: 2000–2010. *Computers, Environment and Urban Systems* 43: 51–66.

McCann, Ph. and Acs, Z. (2011) Globalization: Countries, Cities and Multinationals. *Regional Studies* 45(1): 17–32.

Mayer, H., Sager, F., Kaufmann D. and Warland, M. (2016) Capital City Dynamics: Linking Regional Innovation Systems, Locational Policies and Policy Regimes. *Cities* 51: 11–20.

Pacione, M. (2009) *Urban Geography: A Global Perspective*. London: Routledge.

Parnreiter, Ch. (2014) Network or Hierarchical Relations? A Plea for Redirecting Attention to the Control Functions of Global Cities. *Tijdschrift voor Economische en Sociale Geografie* 105(4): 398–411.

Rérat, P. (2012) The New Demographic Growth of Cities: The Case of Reurbanisation in Switzerland. *Urban Studies* 49(5): 1107–1125.

Robinson, J. (2005) Urban Geography: World Cities, or a World of Cities. *Progress in Human Geography* 29(6): 757–765.

Sassen, S. (1991) *The Global City: New York, London, Tokyo*. Princeton, NJ: Princeton University Press.

Sassen, S. (2007) *A Sociology of Globalization*. New York: W.W. Norton & Co.

Smith, A. and Timár, J. (2010) Uneven Transformations: Space, Economy and Society 20 Years after the Collapse of State Socialism. *European Urban and Regional Studies* 17(2): 115–125.

Sýkora, L. and Bouzarovski, S. (2012) Multiple Transformations: Conceptualising the Post-communist Urban Transition, *Urban Studies*, 49(1): 43–60.

Sýkora, L. and Stanilov, K. (2014) The Challenge of Postsocialist Suburbanization, in K. Stanilov and L. Sýkora (eds), *Confronting Suburbanization: Urban Decentralization in Postsocialist Central and Eastern Europe*. Chichester: Wiley Blackwell.

Taylor, P.J. (1993) *Political Geography: World-Economy, Nation-State and Locality*. London: Longman.

Taylor, P.J. (2000) World Cities and Territorial States under Conditions of Contemporary Globalization. *Political Geography* 19: 5–32.

Taylor, P.J. (2001) Specification of the World City Network. *Geographical Analysis* 33: 181–194.

Taylor, P.J. (2004) *World City Network: A Global Urban Analysis*. London: Routledge.

Taylor, P.J. and Csomós, G. (2012) Cities as Control and Command Centres: Analysis and Interpretation. *Cities* 29: 408–411.

Taylor, P.J. and Derudder, B. (2014) Tales of Two Cities: Political Capitals and Economic Centres in the World City Network. *Glocalism: Journal Of Culture, Politics And Innovation*. Published online by 'Globus et Locus'. Retrieved 17 May 2016 from www.glocalismjournal.net.

Taylor, P.J. and Derudder, B. (2016) *World City Network: A Global Urban Analysis*. London: Routledge.

United Nations, Department of Economic and Social Affairs, Population Division (2015) *World Population Prospects 2015*. Retrieved 15 June 2016 from https://esa.un.org/unpd/wpp/Publications/.

Wolman, H. and McManmon R. (2012) What Cities Do: How Much Does Urban Policy Matter?, in K. Mossberger, E.S. Clarke and P. John (eds), *The Oxford Handbook of Urban Politics*. Oxford University Press, 415–441.

Zhou, Y. and Ma, L.J.C. (2000) Economic Restructuring and Suburbanization in China. *Urban Geography* 21(3): 205–236.

5   Geographies of transportation in the twenty-first century

Gábor Szalkai, Attila Jancsovics, Imre Keserü, Cathy Macharis, Balázs Németh and Vilmos Oszter

Introduction

Transportation, like movement in animal life, is an integral part of human society. Although the rise of information and communication technologies (ICT) in several areas, particularly in social communications and the tertiary sector, has partly replaced transportation, the latter continues to form the basis of the territorial division of labour in agriculture, industry and trade. Due to this essential role, transportation has always been one of the leading research topics in geography.

The state of development of society always correlates strongly with transportation infrastructure, which in turn is determined by constantly developing technology. Mass demand for transport was first satisfied by ship transport, which together with road transport formed the only modes of transportation. These were partly supplemented, and partly replaced, by rail transport in the nineteenth century, but by the end of the twentieth century, transportation was dominated by road transport and aviation.

Maritime transport

Maritime transport, the most important mode of transportation before the Industrial Age, has greatly declined in dominance. Long-distance passenger shipping only exists today due to tourism, while freight shipping, whose only rival in the past was road transport, now faces the challenges of rail and air transport. Despite the emergence of a global economic system, enabled by the development of technology, sea shipping has remained a major subsector, holding an 80 per cent share of the global market (UNCTAD 2013).

Therefore, sea shipping continues to be an economically strong sector, on which other economic activities depend. Shipping is connected to globally generated GDP through shipbuilding, ship demolition, component manufacturing, maritime insurance, fishing, tourism and energy sector. This is the main reason for the fact that the steadily growing quantity of products had decreased only in 2009 due to the global crises, but started to increase from the following year (UNCTAD 2015).

Containerisation, oil transportation, bulk goods (correlating with the growth of population and raw material demand in the developing countries), the growing demand for iron ore, coal, grain, bauxite, alumina and phosphate, the liberalisation of trade and the growing role of the private sector, together with geopolitical changes have all boosted maritime transport in the last few decades. Developing countries have played an increasingly important role in this growth. They now are not only the world's largest exporters but also its most important importers. This is a major change, as developing countries previously functioned primarily as exporters of raw materials. As a result, Asia now leads in sea shipping, followed by Europe, America and Africa (UNCTAD 2013). While the share of developing countries in the volume of exported products is around 20 per cent (and has been unchanged since 1970), in terms of imported products, the figure has gone up from 20 per cent to 61 per cent. Asia has been the catalyst of these changes (UNCTAD 2015).

The geographic location of major ports also illustrates this process, for apart from Rotterdam (eighth), only Asian ports are among the top ten. Only one (Singapore) of the remaining nine is situated outside of China (Geohive n.d.). In terms of shipping fleets, the picture is different: Greece is ranked first, followed by Japan, China, Germany and Singapore. These five countries share more than half of the global merchant navy capacity.

The trade and routes of products transported by sea are affected by the locations of the main occurrences and processing hubs. The Middle East as the main exporter, the drilling of new wells and increasing consumption in India and China defines the conditions for the bulk transport of oil. The increase (or decrease) in import volume of other stakeholders, the growing need for oil reserves and expanding refining capacity are also playing a part in the current process. The Middle East is main exporter of refined products, while South America and developing Asian countries (excluding China) have the largest imports. The rising South–South traffic and the variance from the classic trade connections are redistributing the trade of oil tankers and modifying the main routes of maritime transport (UNCTAD 2013).

Japan and South Korea are the main importers of liquefied natural gas (LNG) (with Australia as the leading exporter), and according to forecasts by 2020, the volume of transported LNG will be doubled. Environmental regulations and emission controls are also factors in the rising importance of gas.

The volume of bulk commodities traffic is also determined by developing countries, particularly China and India. The majority of iron ore is exported from Australia, which with Brazil accounts for 75 per cent of total world exports, while China imports 68 per cent. China is the biggest coal importer, while Indonesia and Australia account for 65 per cent of exports. For all other bulk commodities, the USA is the biggest exporter and Asia is the main importer.

Apart from bulk products, container traffic is the other key segment for maritime transport. Global container transport increased from 50 million TEU (twenty-foot equivalent unit: it measures the cargo capacity of container ships and container terminals) in 1996 to approximately 180 TEU in 2015, experiencing a decline only in 2009. Due to rising traffic volume, the dimensions of container ships are also growing, and the shipping prices decreasing, however, the squeezing out of small enterprises is creating an oligopolistic market, which will lead to increasing prices in the long term. The rising quantity of products to be transported by containers (for instance agricultural products), as well as restructuring in China (with its increasingly important position in the supply chain) are the most significant drivers of the increase in container traffic (UNCTAD 2013).

This process can be seen in the distribution of the most important container terminals. In 2014, the ten terminals with the biggest turnover were all situated in Asia, six in China, including the world's leading terminal in Shanghai. Hong Kong is the most saturated area, where three ports account for more than 9 per cent of global trade. The top non-Asian port, Rotterdam, was ranked eleventh.

The main West–East container traffic routes have lost their significance in recent years, whereas others have grown dynamically. Asia–Europe and Asia–America are the most important West–East routes, while transatlantic shipping has stagnated over the last ten years. Whereas in 1995 transatlantic container trade gave 22 per cent of these directions, by 2014 the figure decreased to 13 per cent due to both America and Europe strengthening their trade connections with the most dynamically developing Asian countries, in particular China in recent years (UNCTAD 2015).

Regarding the future, researchers expect trade, consumption and emissions to grow significantly. The recent completion of the new lock of the Panama Canal is also contributing to this trend, as it allows transit between Asia and the East Coast of the United States. After the enlargement, the canal has three times greater cargo capacity than the previous locks were capable of handling (UNCTAD 2013).

Waterbourne transport continues to be the most environmentally friendly mode of transportation with a CO_2 emission level which is the same as that of railway transport, a fifth of that of road transport and just one-fifteenth of airborne transport emissions. In 2012, maritime transport was responsible for only 2.2 per cent of global CO_2 emissions, but this is expected to increase by 50–250 per cent by 2050 (UNCTAD 2015).

Despite its relatively low environmental impact, sea shipping together with the whole transport sector is currently under political pressure to shift towards more environmentally friendly technologies. Ships fuelled by liquefied natural gas have already appeared on the market, and they may well represent the transport of tomorrow (Berg 2016).

Rail transport

Rail transport is a key inland means of transportation. It has the capacity to transport passengers and goods on a considerable scale and in an environmentally and financially sustainable way.

Railway networks developed from the early nineteenth century onwards around the globe. From its beginnings, rail was a direct, reliable and fast challenge to other means of transportation. After nearly 200 years of continuous technological development, it is possible today to build railway infrastructure in virtually any kind of natural environment regardless of whether it is extremely high mountainous terrain (e.g. the Tibetan Plateau) or a seismologically sensitive deep sea floor (e.g. the Seikan tunnel). Nevertheless, due to economic factors, largely relating to expected traffic demand and the constant development of other means of transport, a comprehensive economic and social analysis must be carried out when a decision on further railway transport infrastructure development is being taken (Blainey, Hickford and Preston 2012). High capacity, efficient railways are costly to build and maintain, but different business (management) models exist around the globe suited to local characteristics (Laurino, Ramella and Beria 2015).

Railway transport services can be divided into two main groups: passenger and freight operations. They are usually separated completely, although the use of mixed trains still exists in Switzerland where they are employed to improve secondary-line efficiency, as well as in a number of peripheral developing countries, where the lack of sufficient infrastructure or rolling stock makes the mixed train the only suitable local rail transport solution. Some of the most important railway networks are freight oriented, including nearly all of North and South America, from Canada and the USA to Brazil and Argentina. In most parts of Africa and Australia, the railway systems are also freight dominated. The role of passenger services is minor in these countries, if they exist at all. In most cases, only a few suburban passenger services operate around bigger cities, together with a number of long-distance intercity trains which can offer competitive journey times compared to other means of transportation. Apart from some suburban areas, passenger services share track with mostly privately owned freight operators where the former have no priority, thus average delays are relatively high for long-distance services (Amtrak 2016). The main geographical factors which favour the domination of freight-oriented railway systems are relatively large average (over 800–1,000 km) transport distances and low population densities combined with high urbanisation rates and sizeable metropolitan areas.

In most European countries and in Japan, the railway networks are increasingly passenger oriented, despite policy efforts made in the past decades. Particularly in the most densely populated Western European countries, the relative short travel distances and the extensive network of overlapping functional urban areas create a stable constant demand for frequent passenger services. Under the current taxation and social welfare systems, and except for a few busy routes, strong travel demand alone is not enough to allow profit-making operations. Thus, essential passenger services have to be co-financed under a PSO (Public Service Obligation) arrangement in which a competent national or regional body orders passenger services (European Commission 2014) from a service provider company. The mostly standardised PSO arrangements defines the exact service offer parameters and usually includes bonus/malus financial incentives for the service provider in order to increase service quality through the lifetime (usually fifteen to twenty years) of the PSO arrangements. Higher standards of living generate increased daily travel needs but due to high car ownership rates, the modal share of railway services are constant or slowly growing from a low base (5 per cent of total trips). As European society ages, its changing lifestyle is slowly shifting the main purpose of journeys from commuting to leisure and health care related. The younger generation is also more mobile than in the past but it has fewer car owners (Kuhnimhof *et al*. 2012) and expects to be served by more door-to-door multimodal transport services which are supported by the latest real-time ITS (Intelligent Transport System), for instance smartphone travel planner applications, real-time passenger information panels, etc. Due to these trends, the demand for passenger railway services is expected to remain the same or may also slowly grow even in countries with shrinking populations. Railway freight transport in Europe is generally under 20 per cent of total transported goods, measured in ton-km (Eurostat 2015b). Unlike in America, the distances are smaller and due to different historical development, the destinations are more diverse in space than in the more concentrated American urban structures. The flexibility of the strong road competition supported by several policy instruments which favour the road sector (e.g. low road infrastructure usage fees, lack of external cost taxation, etc.), mean that railway freight services are generally not economically attractive transport alternatives. Furthermore, European freight trains as a rule have to share track with relatively frequent passenger trains to which they nearly always have to give way, which makes freight services slow and unsuited to most high-tech industries' just-in-time transport needs. As far as Japan is concerned, it is worth noting that coastal shipping is the main competitor for railway freight as nearly all the important urban areas and industries are spread along the coast.

Two of the world's railway systems – China, Russia – which both have huge passenger and freight operations can be characterised as mixed, where both passenger and freight services have an important role. Their growing, prosperous railway systems have to cope with growing population numbers (particularly in urban areas) and the need for more freight transport due to demand from developing industries and logistics activities. Besides some urban areas, both countries lack comprehensive high capacity motorway networks while inland or coastal waterways are not comparable alternatives to railway freight, nor are their non- or only partly liberalised air transport industries. China has the biggest population and greatest economic power of the two with a balanced urban network located in the Central Eastern part of its territory (Wu, Nash and Wang 2014). High population density and rapidly growing urban areas with very high network utilisation make China's railway system one the most intensively used in the world. Both technical and network development (high-speed and conventional) have been extremely fast. In a less than ten-year period, China has built the longest high-speed railway network in the world (Shaw, Fang, Lu and Tao 2014) which by 2015 (according to UIC) alone carried roughly the same number of passengers (800 million) as all other such networks together. Combined with the huge population concentrations, the fact that for door-to-door journeys of up to 1,000 km or five hours, high speed trains provide the shortest and most reliable journey times ensures the sustainability of the long-term high-speed railway network development aims.

Russian Railways (RZD) are gradually positioning themselves towards a more freight-oriented system as the population density in most regions of Russia is extremely low and only along certain (rail) corridors is there any significant travel demand between cities. In the Asian part of Russia, the Trans-Siberian Railway has become a major factor for the development of the urban structure in areas through which it runs. For journeys over 1,000 km, aeroplanes play the leading role, while in the bigger suburban areas, distances of under 100 km are dominated by cars and buses. Nevertheless, the gradually developing Asian–European land bridge, containerised traffic and the huge volume of commodities and petroleum products will likely secure the long-term future of railway freight in Russia (Polikarpov 2015).

Provided that there is sufficient travel demand along the railway lines which can be stimulated by efficient multimodal passenger and combined freight services, rail is the most sustainable universal inland means of transport, excepting the slower and more territorially restricted inland waterways. The emission and noise levels per passenger and ton-km of rail transport are significantly lower than those of its main competitors – road and air. Different forms of high capacity urban and suburban railway networks are the main transport solutions for rapidly growing global megacities, and the increasingly urbanized developing countries with their huge mobility needs. Decision makers should, through policy measures, put more emphasis globally on ensuring fair competition and an operational environment for sustainable railway services.

Road transportation

Without doubt, today's most crucial energy source is oil. This is especially true in the transportation sector, as 93 per cent of the sector's energy consumption comes from petroleum (and petroleum products) (IEA).

Apart from oil's economic disadvantages (dependency, uncertain prices), pollution may be an even greater issue. Carbon dioxide as the most important green house gas (representing 76 per cent of global GHG emissions) is just one contributor to the transport sector's air pollution, with 23 per cent of global CO_2 emissions coming from fuel combustion (IEA 2015). However, other pollutants from combustion engines pose very serious health issues, primarily photochemical smog (ground-level ozone) and atmospheric particulate matter (PM). This is particularly severe in urban areas, where emissions from combustion engines are at their highest concentration (due to high traffic volumes). The air conditions of, for example, Beijing, Delhi, Mexico City and Los Angeles serve as an excellent indicator of this phenomenon (WHO 2016).

Although the current problems are serious enough, in the upcoming decades, they are expected to become even more severe as the number of cars and other vehicles continues to grow rapidly due to the economic development of China, India, Indonesia and other developing countries. The year 2010 was the first in history when the global number of motorized vehicles in operation surpassed 1 billion (Wardsauto n.d.) and by 2014, it reached 1.2 billion (OICA n.d.). This increase does not seem to be slowing down, as between 2005 and 2014 the motorization rate increased globally by 25 per cent, reaching 180 vehicles/1,000 inhabitants (and also in light of the continuing rise in global population). The biggest increase in this period occurred, unsurprisingly, in Asia, which saw an overall increase of 123 per cent, including China with 102 per cent and Southeast Asia with 232 per cent (OICA n.d.). According to some scenarios, by the year 2030, more than 2 billion motorized vehicles could be on the roads.

Currently, electric engines seem to be the only serious, potential alternative technology to regular internal combustion engines (ICEs), as neither biofuels (due to the sheer lack of enough arable land on the planet) nor hydrogen cells hold much promise.

In terms of pollution issues, electric engines have no direct emissions at all, which is a really critical point in already heavily polluted urban areas. The source of the electricity which drives these engines is of course an important matter, but even with the current mix of power stations, the total CO_2 output is lower for currently existing electric vehicles than combustion engine cars (Union of Concerned Scientists 2015). A further advantage of electric engines is their part in eliminating other pollutants like particular matter and smog in densely populated areas.

In 2015, the global electric car stock reached 1.26 million (OECD/IEA 2016), which was more than 100 times larger than the 2010 estimate. This marked the first time the threshold of 1 million electric cars on the road had been exceeded (including both Plug-In Hybrid Electric Vehicles [PHEV] and full electric or Battery Electric Vehicles [BEV]). In 2016, another great threshold was surpassed as more than 2 million electric cars were on the road globally. Also, China, overtaking the USA, has become the country with the largest – about one-third of the global – electric car stock. With more than 200 million electric two-wheelers, about 4 million low-speed electric vehicles (LSEVs) and with more than 300,000 electric buses, China is clearly the global leader in the electrification of transportation.

The global electric car stock growth exceeded 59 per cent in 2016, 77 per cent in 2015 and 84 per cent in 2014, slightly declining from the period 2011–2013 when the number of electric cars more than doubled every year. In 2014 and 2015, the strongest increases took place in China, Korea, the United Kingdom, Sweden, Norway, the Netherlands and Germany (OECD/IEA 2017).

Despite the previous anti-electric policies of car manufacturers, the newcomer Tesla Motors was able to serve as a real catalyst for the electrification of the car industry by creating a competitive model of electric car. It is a remarkable fact in itself that a new car company could gain entry to a market dominated by some of the world's greatest firms with decades of experience, technology and infrastructure. Especially noteworthy was Tesla's 2014 announcement that it was making the company's electric car technology patents available to all, even their competitors, with the goal of speeding up innovation and stimulating further growth in the EV industry.

Not just the release of patents, but the mere existence of electric engine technology has long been considered a threat to traditional car companies. These, no doubt reasonably, feared the loss of their incumbency-based advantages in combustion engine research and manufacture (which prevented newcomers appearing in the first place), due to the development of electric motors and electric power trains. As a result the century-old business model of car manufacturing would be shattered to pieces in the upcoming decades, as now even the incumbents' suppliers would be able to research and sell their own products (as has already happened).

Presumably realizing the inevitable future electrification of the automobile industry, several large traditional car companies have put great efforts over recent years into EV research and have produced an increasing number of EV models. One of the biggest players, Volkswagen, which was severely affected by its 2015 emissions scandal, perhaps as a comeback strategy, has announced the introduction of thirty new EV models by 2025, and in 2016, its target of selling 2–3 million electric cars by 2025. Ford announced thirteen new models by 2020, Honda wishes to produces two-thirds of its model palette with EVs by 2030 and Tesla aims to sell 1 million cars by 2020 (OECD/IEA 2017; OECD/IEA 2016).

General Motors (which has had its own [unpleasant] experience of electric cars with the 1996 EV1 model) launched its BEV model Chevy Bolt in December 2016. As was announced earlier, it has a range of about 383 km (combined urban and highway driving), and current [2017] prices are around US$37,500, which after various government and local tax reliefs could mean a final cost of just under US$30,000 (Chevrolet n.d.). Based on the experiences of the company's first 'mass-market' electric car, in October 2017, General Motors announced plans to introduce two new electric models within eighteen months, and deliver at least twenty new electric vehicle models globally by 2023 (Chevrolet n.d.; Lambert 2017).

In spite of the current fast-evolving EV trends, the global market share of electric cars is still very low (0.1 per cent in 2015) compared to the total number of passenger cars on the road worldwide, which was close to 1.2 billion in 2015 (OECD/IEA 2016; Wardsauto n.d.). However, several predictions expect EV stock to reach 20 million by 2020 (1.7 per cent share) and 100 million by 2030.

Moreover, during the last few years, we have witnessed unprecedented support from governments and regulators for electric cars, which has enabled them to gain the momentum needed to replace combustion engine cars far sooner than was previous predictions suggested. China, the largest car manufacturer of the world, has announced in 2017 that will join with the UK and France to ban ICEs selling by 2030, and completely remove them from the roads by 2040. Furthermore, Paris announced in 2017 that it wishes to entirely ban ICEs by 2030 while London, Barcelona and even Stuttgart, home of the Mercedes-Benz have since also announced similar initiatives (Beene and Lippert 2017; Smith 2017).

These assisting actions could greatly push forward electrification, so much that current predictions of EV stocks could be as underestimating as they were a few years ago.

It is no surprise that batteries are the alpha and the omega of EV technology, as they also represent its greatest weak point. One key factor is the maximum driving range (which has evolved significantly over the years, and in some of the newest models it even surpasses 450 km), while recharge time is another important element. In relation to charging infrastructure (it should be noted that in most countries recharge stations are free), a large number of chargers would be necessary for a reasonably sized EV car park. The Tesla Supercharger station network (which, in September 2016, consisted of 705 stations and 4,359 chargers globally, growing to 1,032 stations and 7,320 charges by October of 2017, and planning to install 1,000 more chargers in China by the end of 2017) allows the Model S to charge 270 km of energy within thirty minutes and to fully charge in under seventy-five minutes (Teslamotors n.d.; Tesla n.d.). Even so a record was set with a Model S100D this year, travelling for 1,072 km before stopping, with an average speed of 40 km/h (Gibbs 2017).

Clearly, the biggest disadvantage of electric cars is their overall high prices, and this is the direct result of expensive batteries. Therefore, it is critical to increase battery energy density and reduce costs. Fortunately, recent trends in this area are positive and it seems likely that targets set by car manufacturers will be met (Figure 5.1).

However, a breakthrough in battery technology could completely revolutionize not just the electric car industry but also several other battery-related industries such as smart phones, tablets, laptops and renewable energy power stations. Furthermore, a technological discovery coming from any of these areas could

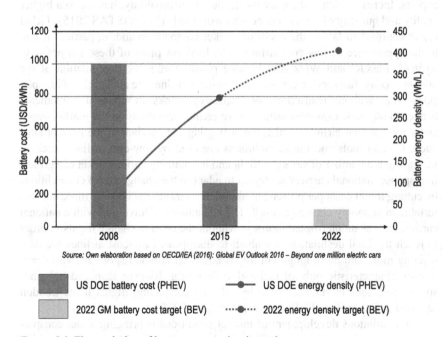

Source: Own elaboration based on OECD/IEA (2016): Global EV Outlook 2016 – Beyond one million electric cars

Figure 5.1 The evolution of battery energy density and cost.

have a big impact, and perhaps even solve the problem of renewable energy storage and battery life at one stroke. It is small wonder that this area is an important focal point of research.

A major improvement in battery capacity could quickly lead to the revision of previous forecasts for EV stocks. Moreover, if all the disadvantages which derive from batteries are eliminated, combustion engine cars could become far less competitive in relation to EVs.

The coming decades are faced with the challenge of satisfying the increasing demand for mobility and, therefore, maintaining a sustainable global car stock. This raises the crucial question of whether electric cars can provide a solution. Currently, the signs are very promising and just getting better and better.

Air transportation

Only two decades ago, before the new millennium, air transportation had a rather different meaning for people than today. In the twenty-first century, aviation went through significant changes in terms of availability, affordability and the most visible factor: connectivity. In today's world, air transportation has become easily accessible in the developed parts of the world thanks to the liberalization of the entire market between the 1970s and 1990s. Changes in aviation policy opened many new doors for the development of air travel, and different parts of the world were linked to each other by faster and more frequent flights. Deregulation of the sector directly contributed to increases in GDP, employment, travel, tourism and exports. Increased demand for air travel, due to affordability, has led to a higher quality and quantity of direct connections worldwide (InterVISTAS 2015). Ticket prices continue to fall as the result of market competition and, in particular, due to the appearance of low-cost airlines. The business plans of these airlines (e.g. Ryanair, EasyJet and Wizz Air) became possible after the deregulation of the market. Today, full service carriers and low-cost airlines are able to fly 'with open skies' (i.e. without restrictions on capacity) thanks to bilateral agreements between countries. Low-cost airlines have created instability in the market, especially for national airlines, which are struggling for survival in the current situation. Certain tools and market solutions are used by low-cost airlines, such as increasing the number of seats per flight and introducing various ticket categories. In response, national carriers are trying to adapt to the changed market conditions by cutting travel costs per passenger, although air travel has become more uncomfortable in economy class as a result. The advantages of travelling with a national carrier, such as uninterrupted journeys without the need to change flights in order to reach the final destination, are about to disappear. Low-cost airlines are now offering more and more long-haul flights based on hub-and-spoke networks previously characteristic only of national airlines (e.g. Ryanair launched hub-and-spoke operations in summer 2016 by offering connecting flights in its London Stansted and Barcelona El-Prat bases) (Mulligan 2016).

The continuous development of market conditions is bringing about conspicuous changes in market and network shares globally. Moreover, emerging markets like India, China and the countries of South-East Asia are also reshaping the world's air transportation network. According to a 2016 report from IATA, India's domestic air passenger traffic had increased by 21 per cent year-on-year in May 2016 (while increases in the US and Australian markets were 4.4 per cent and 0.8 per cent, respectively) (IATA 2016). In order to catch up with the increasing demand in these regions, networks have to expand, and connectivity has to develop between the core regions (North America and Europe) and newly emerging ones. Based on current global trends, 11–25 per cent of European demand for air transport by 2030 will be unmet, with twenty to forty major European airports expected to be chronically overcrowded due to increased connections (ACI Europe 2010). Until the late 2000s, European and American airports dominated the list of the busiest and most crowded airports in the world in terms of both passenger and freight numbers. However, more recently, major airports in the Middle East have experienced unprecedented development in terms of connectivity and passenger numbers. This process is related to the appearance of emerging markets in the air transportation network as these airports now have the best geographical conditions to connect existing hubs with emerging ones. National airlines in the Middle East, such as Emirates and Qatar Airways are taking advantage of these opportunities with major new developments to expand their capacity and networks to deal with growing passenger numbers. As already mentioned before, major European airports are currently facing problems with a lack of capacity and a shortage of time slots. Therefore, airports in the Middle East are filling this gap in the market by rapidly increasing airport connectivity in order to connect North America and Europe with the emerging markets. According to a report in 2004, European hub airports had three and a half times more international connections than airports in the Middle East. This proportion has now reversed as in 2014 Middle Eastern airports served twice as many connections as their European counterparts (ACI Europe 2014).

As human space has broadened as a result of globalization, growing demand requires airline network expansions and increases in flight frequency. Average flight numbers per day exceeded 100,000 in 2014 and continue to grow by 2–3 per cent annually (ATAG 2014). Air travel has a huge carbon footprint due to its enormous fuel emissions. It is responsible for 2 per cent of the world's human-induced carbon dioxide emissions and 12 per cent of emissions from all transport sources (ATAG 2014). Consequently, aviation has to evolve to become more environmentally friendly in the next decades. According to IATA, commercial aviation will grow by 5 per cent annually until 2030, while fuel efficiency improvements will develop only by 3 per cent (IATA 2011). This means that existing emissions will continue to grow if there is no development in fuel efficiency and its related technology, and in the type of fuel used. Alternative fuels like sustainable biofuels began to appear on the market in 2011 and some airlines (e.g. United, KLM, Lufthansa and Qantas) are already using biofuels in live operations. More than 2,200 commercial flights have flown on biofuels since 2011, and this number is expected to grow steadily in the coming years (ATAG 2016).

Box 5.1 Dubai Airports

A good example of this remarkable development process is the case of Dubai Airports. The rapid urban development of this Middle Eastern metropolis is widely familiar, but the city is also currently experiencing conspicuous development in terms of air transportation. This is due to the fact that 'over 2/3 of the world's population lives within 8 hours flight time from Dubai' (Dubai Airports 2013: 14). The city had one main airport (Dubai International Airport – DXB) until 2013, when a new 'airport city' Dubai World Central – DWC opened its doors to passengers and freight. The older airport itself is now one of the busiest airports in the world, with more than 78 million passengers annually and connections to 260 destinations (Dubai Airports 2016). Moreover, it is forecast to reach 100 million passengers by the end of the decade. Dubai World Central is an entire airport city (including El Maktoum International Airport), which gives an impression of what airports of the future will look like with all kinds of business and leisure facilities available in one place. Once the entire project is finished, DWC will be the world's largest global airport hub with an expected 160 million passengers passing through it annually by 2050 (Dubai Airports 2014) (Figure 5.2). It will also serve a multimodal logistics hub for 12 million tonnes of freight. Dubai's proximity to emerging markets such as India and China and its geocentric location between continents further elevates its status as a global centre for trade, tourism and business (Dubai Airports 2013).

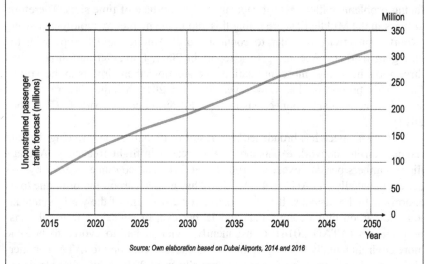

Source: Own elaboration based on Dubai Airports, 2014 and 2016

Figure 5.2 Traffic predictions to 2050 for DXB and DWC combined.

The IATA is committed to reaching carbon neutral growth by 2020 and a 50 per cent reduction in carbon dioxide emissions by 2050 (IATA 2013). In order to achieve these targets, technological developments like fuel efficiency have to continue, combined with a gradual increase in the use of biofuels. Over their full life cycle, biofuels derived from biomass have been shown to reduce the carbon footprint of aviation fuel by up to 80 per cent (ATAG 2016).

The future of air transportation is hard to predict based on its dynamic development over the past few decades, but a few things can be inferred from current technological developments and trends. Tests of electric powered and solar impulse flights are appearing in news feeds daily and supersonic flights are expected to come back into operation by 2025, with hydrogen powered flights predicted by 2040. In just a few years' time, a passenger will perhaps be able to buy a $15 ticket for a transatlantic flight operated by a low-cost airline, and 30 per cent of these flights will be powered with biofuel by 2030 (BBC Future 2015).

Current trends in urban mobility

Urban areas are increasingly important in terms of the global challenges that affect the transport sector. In 2014, 54 per cent of the world's population lived in cities. The United Nations (2015a) estimates that this proportion will increase to 66 per cent by 2050. Urban mobility is estimated to account for 23 per cent of all CO_2 emissions in Europe (European Commission 2016) and transport will be the largest source of carbon emissions in Europe by 2030 (Capros, De Vita, Tasios, Papadopoulos and Siskos 2013). Urban transport is also responsible for a significant amount of pollution in cities. According to the European Environmental Agency, 96 per cent of city dwellers were exposed to harmful particulate matter concentration ($PM_{2.5}$) between 2009 and 2011 and urban transport contributes significantly to PM_{10} emissions, NO_2 concentrations and noise. The land take of transport infrastructure also influences the quality of life in urban areas, contributes to irreversible soil sealing and the loss of green space (European Environmental Agency 2013). Road traffic congestion, now an everyday phenomenon in major urban areas, is a major cause of economic loss due to time wasted in traffic jams (Figure 5.3).

The above-mentioned problems in contemporary cities are the result of the 'predict and provide' planning paradigm characteristic of the twentieth century, which focused on satisfying an ever-increasing demand for road capacity and smooth flow of car traffic, and resulted in primarily car-oriented cities. By the 1970s, it had become clear that the uncontrolled growth of urban traffic had caused numerous problems affecting the quality of life in cities (e.g. environmental pollution, safety). Therefore, the new discourse of transport planning which has since emerged in the Western world focuses on the management of travel demand and the promotion of a modal shift from car transport to sustainable modes such as public transport, cycling and walking. As a result of this paradigm change, even in the United States, where the car is deeply entrenched

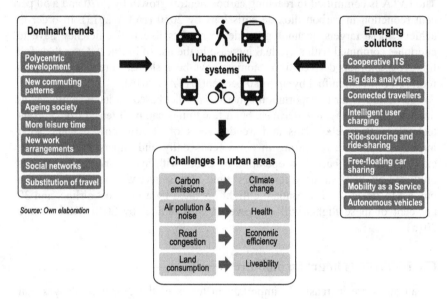

Figure 5.3 Current challenges, trends and solutions in urban mobility.

in people's daily mobility, public transport use increased by 25 per cent between 1995–2015, exceeding population growth (American Public Transport Association 2016). In addition, many Western countries have reported an end to rises in car ownership and distances travelled by car (see Box 5.2).

The efforts of Western countries to mitigate CO_2 emissions and limit energy use are, however, offset by the rapidly increasing pollution and fossil energy use in emerging economies (Tran 2013). Although car ownership rates are still under 200 cars/1,000 capita even in the most developed cities of emerging economies (Woldeamanuel 2016), some countries like China show a very fast rate of increase. Wang, Teter and Sperling (2011) estimate that the growth rate of car ownership in China will reach 13–17 per cent per year with enormous negative consequences for global carbon emissions and fossil energy use.

While there is consensus about the need to make urban mobility more sustainable, a number of recent spatial, demographic and lifestyle trends pose new challenges to urban mobility systems (Figure 5.3).

One of the main emerging spatial trends is the increasing polycentricity of urban areas with the major origins and destinations of travel (e.g. home, work, shopping and leisure) dispersed in metropolitan space. As a result, new commuting patterns such as reverse commuting from the central city to the suburbs and cross-commuting within the suburban areas have emerged (Small and Verhoef 2007; van der Laan 1996). These changes have induced a transformation in travel behaviour, with increasing trip lengths and durations, and a mishmash of commuting routes, which are more difficult for high-capacity, rail-based public

---

Box 5.2 Is the continuous growth of car traffic over? The peak car hypothesis

Until the mid-2000s, car use measured by car ownership and distance travelled was continuously increasing in the Western world. There is, however, growing evidence that the average annual distance travelled has now stopped rising. The phenomenon has been coined as 'peak car' and has been observed in most developed countries (Kuhnimhof, Zumkeller and Chlond 2013; Millard-Ball and Schipper 2011). A potential enabler of this change is the emergence of information and communication technologies (ICT), which have made it possible to avoid trips (e.g. thanks to teleworking and online shopping) and enabled new types of mobility such as car sharing. In addition, many urban areas are choosing to promote car-free lifestyles by introducing access limitations, car-free areas, pedestrianization schemes and traffic calming measures, while public transport is increasingly perceived as a realistic, often faster and cheaper, alternative to the car (Goodwin and Van Dender 2013). Another factor is that young people in many countries are less interested in obtaining a driving licence at a young age and starting to drive (Delbosc and Currie 2013). This is especially true for the unmarried and those who live in urban areas with good access to public transport (Hjorthol 2016). Recent trends in urbanization have also been mentioned as causes of the peak car phenomenon. Settlement size, population density, the degree of urban development and the revival of metropolitan centres have been found in recent research to be important determinants of the reduction in car use (Grimal, Collet and Madre 2013; Headicar 2013). The future evolution of modal shares will possibly depend on land use policies for housing and employment designed to cater for the growing urban populations (Metz 2013).

---

transport to serve. The end result has been to shift commuters' mode choice to cars.

Besides changes in the spatial distribution of the population and their activities, changes in the demographic structure of society are also of importance. Ageing is one of the main demographic trends impacting urban transport. In 2015, one in eight people were aged sixty or over in the world. The United Nations projects that by 2030, this will increase to one in six, with a total of 1.4 billion people aged sixty or over, up by 56 per cent on 2015 (United Nations 2015b). Since in most European countries (with the exception of Belgium and Poland) elderly populations are concentrated in rural areas (Eurostat 2015b), transport provision has to be adapted to the needs of the elderly, who may be more reliant on public transport services and require special facilities such as low-floor vehicles, wheelchair access and financial subsidies.

Future travel demand will also be determined by an ever-increasing mix of interrelated lifestyle factors such as changes in how people spend their time (more leisure time), flexible work arrangements (e.g. telework), new social networks (e.g. through social media) affecting the demand for travel and technologies that may substitute for travel (e.g. virtual reality).

The range of technological and organisational solutions for the management of urban mobility systems and improvement of their sustainability is quickly expanding due to the rapid development of information and communication technologies (ICT) (Figure 5.3). Digital data collectors such as traffic monitoring cameras, public transport smartcards, 'internet of things' devices, built-in vehicle sensors and smartphones produce enormous amounts of data and enable the introduction of Cooperative Intelligent Transport Systems (C-ITS). Such systems are based on wireless digital communication connecting vehicles with each other and the transport infrastructure. They facilitate real-time multimodal traffic management to increase safety, avoid bottlenecks, improve information provision and promote multimodal travel (Picone, Busanelli, Amoretti, Zanichelli and Ferrari 2015). Big data analytics[1] enables real-time traffic management and demand forecasting. Digital information provision has enabled the emergence of the 'connected traveller', who can independently make decisions about travel routes, modes and departure times based on multimodal route planning, and who receives real-time information about traffic disruptions and gives instant feedback about service quality or problems through connected mobile devices. Furthermore, digitalization and big data analytics could expand the ability of local authorities to introduce financial measures regulating motorized traffic in city centres by intelligent dynamic road user charging and parking pricing reflective of real-time supply and demand.

Digitalization and the proliferation of ICT have stimulated the emergence of disruptive technologies and business models. Online ride-sourcing (the equivalent of taxis) and ride-sharing (sharing a car with other passengers travelling in the same direction), such as UBER and Lyft, have emerged offering flexibility, transparency, low costs and easy ordering and payment procedures. Ride-sourcing may replace longer public transport trips, but also acts as a complement to public transport. The overall impact on traffic and congestion is, however, still not known (Rayle, Shaheen, Chan, Dai and Cervero 2014).

Another service enabled by digitalization is free-floating car sharing,[2] which could solve the first- and last-mile problem, that is, how to reach public transport terminals (Firnkorn 2012). Shared cars may reduce parking demand in cities in the event that they lead to a fall in private car ownership (Firnkorn and Müller 2015).

Digital technologies also make it possible to better integrate transport modes. The concept, known as Mobility as a Service (MaaS), would integrate all public, private and shared travel modes and provide a one-stop-shop for route planning, booking, payment and navigation. MaaS would enable on-demand mobility responsive to the level of user demand and user needs in real time. Such systems would promote a shift from the ownership of vehicles to the use of mobility services, and so may also contribute to a reduction in car use.

Autonomous vehicles are expected to be 'game changers' in future mobility systems. It is estimated that by 2022–2025 autonomous cars will be commercially available with human assistance (an operator will still need to sit behind the controls) (Fagnant and Kockelman 2015). In one or two decades after this, autonomous, driverless, free-floating car-sharing services may become available bringing a paradigm change and with it a potential end to the domination of private cars in cities (Firnkorn and Müller 2015). One probable impact is a significant decrease in parking demand. Zhang, Guhathakurta, Fang and Zhang (2015) showed that up to 90 per cent of parking demand could be eliminated by the introduction of shared autonomous transport systems. The space freed up could be used to improve the quality of public space which, in turn, would make walking and cycling more desirable.

There is, however, a risk that disruptive technologies such as on-demand intelligent mobility (e.g. ride-share services), vehicle automation and electrification of road transport will actually generate more demand for mobility by providing increasingly widespread access to these services and vehicles, and at the same time improve the 'image' of individual travel. Automated vehicles, for example, may enable those who are too old or young to drive to become more mobile. Empty runs, for example, to pick up the owner or park the car in a less expensive area and the potential proliferation of easy-to-use automated car- and ride-sharing services may actually increase demand for road travel and contribute to the maintenance of a car-oriented culture. The possibility of spending travel time efficiently, for example, in an automated vehicle, may increase trip times and distances and therefore support urban sprawl (Levinson 2015). Nevertheless, some of these negative impacts may be mitigated by increasing road capacity due to better traffic management (Fagnant and Kockelman 2015).

Overall, these technological solutions aim to meet current and future travel demand. It is, however, important to mention that other approaches may be just as, if not more, efficient. Thus, land use planning and incentives for behavioural change may bring origins and destinations closer to each other and consequently reduce travel distances. This would make walking and cycling more realistic alternatives to driving and public transport.

Conclusions

This sectorial review confirms that the future of transportation will be determined by at least three major factors. The first, as throughout history, is the current state of technological development. The second, which is a relatively new parameter, is the necessity to reduce the environmental impact of transportation. Finally, spatial, demographic, behavioural and lifestyle trends will continue to determine travel and transport demand as well as the choice of transport mode. The three factors are strongly interrelated. It is expected that alternatively fuelled vehicles will contribute to independence from oil and thus also to the reduction of emissions. Another technological trend is the roll-out and proliferation of automated transport systems and autonomous vehicles, which may revolutionize transport, particularly by road. All these technological changes will, however, not be efficient in solving problems if the demand for travel and transport influenced by socio-demographic trends is not understood properly.

Notes

1 Big data are large datasets, which are analysed by special analytical techniques in order to reveal patterns and trends, especially in relation to human behaviour.
2 A car-sharing scheme where vehicles can be left parked anywhere within a designated zone at the end of the rental as opposed to station-based car sharing, where the vehicle has to be returned to the station it was picked up from.

References

ACI Europe (2010) *An Outlook for Europe's Airports, Facing the Challenges of the 21st Century*. Airports Council International.

ACI Europe (2014) *Airport Industry, Connectivity Report*, 24th ACI Europe Annual Assembly. Retrieved 10 July 2016 from www.seo.nl/uploads/media/ACI_EUROPE_Airport_Connectivity_Report_2004-2014.pdf.

American Public Transport Association (2016) *APTA Ridership Report*. Retrieved 29 July 2016 from www.apta.com/resources/statistics/Pages/ridershipreport.aspx.

Amtrak (2016) Amtrak Monthly Performance Report. Retrieved 25 August 2016 from www.amtrak.com/ccurl/367/406/Amtrak-Monthly-Performance-Report-June-2016.pdf.

ATAG (Air Transport Action Group) (2014) *Aviation Benefits Beyond Borders – Powering Global Economic Growth, Employment, Trade Links, Tourism and Support for Sustainable Development Through Air Transport*. Oxford Economics for ATAG. Retrieved 9 July 2016 from http://aviationbenefits.org/media/26786/ATAG__AviationBenefits 2014_FULL_LowRes.pdf.

ATAG (Air Transport Action Group) (2016a) *Sustainable Fuels – Environmental Efficiency*. Retrieved 10 July 2016 from http://aviationbenefits.org/environmental-efficiency/sustainable-fuels/.

ATAG (Air Transport Action Group) (2016b) *Facts and figures*. Retrieved 10 July 2016 from www.atag.org/facts-and-figures.html.

BBC Future (2015, 10 June) Timeline into the Future of Aviation. Retrieved 9 July 2016 from www.bbc.com/future/sponsored/story/10062015-timeline-into-the-future-of-aviation?ocid=nativedassaultoutbrain.

Beene, R. and Lippert, J. (2017, 26 September) California Considers Following China With Combustion-Engine Car Ban. *Bloomberg*. Retrieved 12 November 2017 from www.bloomberg.com/news/articles/2017-09-26/california-mulls-following-china-with-combustion-engine-car-ban.

Berg, N. (2016) *The Future of Freight: More Shipping, Less Emissions?*. 5 January. Retrieved 1 September 2016 from www.greenbiz.com/article/future-freight-more-shipping-less-emissions.

Blainey, S., Hickford, A. and Preston, J. (2012) 'Barriers to Passenger Rail Use: A Review of the Evidence'. *Transport Reviews* 32(6): 675–696.

Capros, P., De Vita, A., Tasios, N., Papadopoulos, D. and Siskos, P. (2013) *EU Energy, Transport and GHG Emissions Trends to 2050: Reference Scenario 2013*. Luxembourg: European Commission. Retrieved 29 July 2016 from http://ec.europa.eu/transport/media/publications/doc/trends-to-2050-update-2013.pdf.

Chevrolet (n.d.) Retrieved 6 September 2016 from www.chevrolet.com/bolt-ev-electric-vehicle.html.

Delbosc, A. and Currie, G. (2013) Causes of Youth Licensing Decline: A Synthesis of Evidence. *Transport Reviews* 33(3), 271–290. Retrieved 29 July 2016 from http://doi.org/10.1080/01441647.2013.801929.

Dubai Airports (2013) *Strategic Plan 2020 – Connecting the World Today & Tomorrow*. Retrieved 10 July 2016 from www.dubaiairports.ae/docs/default-source/Publications/dubai-airports-strategic-plan-2020.pdf?sfvrsn=2.

Dubai Airports (2014) *The Airport of the Future – Dubai Airports 2050*. Retrieved 10 July 2016 from http://dubaix.ae/wp-content/uploads/2014/04/DXB-2050-eng.pdf.

Dubai Airports (2016) *Fact Sheets*. Retrieved 10 July 2016 from www.dubaiairports.ae/corporate/media-centre/fact-sheets/detail/dubai-airports.

The Economist (2017, 14 September) China Moves Towards Banning the Internal Combustion Engine. Retrieved 12 November 2017 from www.economist.com/news/business/21728980-its-government-developing-plan-phase-out-vehicles-powered-fossil-fuels-china-moves.

European Commission (2014) Commission Staff Working Document – Accompanying the document – Report From the Commission to the Council and the European Parliament. *Fourth report on monitoring development in the rail market*, 2014, [COM(2014) 353 final].

European Commission (2016) *Commission Staff Working Document. Accompanying the document: Communication from the Commission to the European Parliament, The Council, The European Economic and Social Committee and the Committee of the Regions – A European Strategy for Low-Emission Mobility* (No. SWD/2016/0244 final). Brussels: European Commission. Retrieved 29 July 2016 from http://eur-lex.europa.eu/legal-content/EN/TXT/?uri=SWD:2016:244:FIN.

European Environmental Agency (2013) *A Closer Look at Urban Transport: TERM 2013: Transport Indicators Tracking Progress Towards Environmental Targets in Europe*. EEA REPORT, (11/2013). Retrieved 29 July 2016 from https://trid.trb.org/view.aspx?id=1279708.

Eurostat (2015a). *EU Transport in Figures – Statistical Pocketbook*. Retrieved 25 August 2016 from http://ec.europa.eu/transport/facts-fundings/statistics/doc/2015/pocketbook2015.pdf.

Eurostat (2015b) *Statistics on European Cities* (Statistical books). Eurostat. Retrieved 29 July 2016 from http://ec.europa.eu/eurostat/statistics-explained/index.php/Statistics_on_European_cities.

Fagnant, D.J. and Kockelman, K. (2015) Preparing a Nation for Autonomous Vehicles: Opportunities, Barriers and Policy Recommendations. *Transportation Research Part A: Policy and Practice* 77: 167–181. Retrieved 29 July 2016 from http://doi.org/10.1016/j.tra.2015.04.003.

Firnkorn, J. (2012) Triangulation of Two Methods Measuring the Impacts of a Free-Floating Carsharing System in Germany. *Transportation Research Part A: Policy and Practice* 46(10): 1654–1672. Retrieved 29 July 2016 from http://doi.org/10.1016/j.tra.2012.08.003.

Firnkorn, J. and Müller, M. (2015) Free-Floating Electric Carsharing-Fleets in Smart Cities: The Dawning of a Post-Private Car Era in Urban Environments? *Environmental Science & Policy* 45: 30–40. Retrieved 29 July 2016 from http://doi.org/10.1016/j.envsci.2014.09.005.

Geohive (n.d.) Retrieved 1 September 2016 from www.geohive.com/charts/ec_ports.aspx.

Gibbs, S. (2017, 7 August) Tesla Drivers Claim Model S Distance Record of 670 Miles on One Charge. *The Guardian*. Retrieved 12 November 2017 from www.theguardian.com/technology/2017/aug/07/tesla-drivers-claim-model-s-distance-record-of-670-miles-on-one-charge-elon-musk.

Goodwin, P. and Van Dender, K. (2013) 'Peak Car' — Themes and Issues. *Transport Reviews* 33(3): 243–254. Retrieved 29 July 2016 from http://doi.org/10.1080/0144164 7.2013.804133.

Grimal, R., Collet, R. and Madre, J.-L. (2013) Is the Stagnation of Individual Car Travel a General Phenomenon in France?: A Time-Series Analysis by Zone of Residence and Standard of Living. *Transport Reviews* 33(3): 291–309. Retrieved 29 July 2016 from http://doi.org/10.1080/01441647.2013.801930.

Headicar, P. (2013) The Changing Spatial Distribution of the Population in England: Its Nature and Significance for 'Peak Car'. *Transport Reviews* 33(3): 310–324. Retrieved 29 July 2016 from http://doi.org/10.1080/01441647.2013.802751.

Hjorthol, R. (2016) Decreasing Popularity of the Car?: Changes in Driving Licence and Access to a Car Among Young Adults over a 25-Year Period in Norway. *Journal of Transport Geography* 51: 140–146. Retrieved 29 July 2016 from http://doi.org/ 10.1016/j.jtrangeo.2015.12.006.

IATA (2011) *Vision 2050*. Retrieved 10 July 2016 from www.iata.org/pressroom/facts_ figures/Documents/vision-2050.pdf.

IATA (2013) *Reducing Emissions from Aviation Through Carbon-Neutral Growth from 2020*. Retrieved 10 July 2016 from www.iata.org/policy/environment/Documents/atag-paper-on-cng2020-july2013.pdf.

IATA (2016) Air Passenger Market Analysis, May 2016. Retrieved 10 July 2016 from www.iata.org/whatwedo/Documents/economics/passenger-analysis-may-2016.pdf.

IEA (n.d.) Retrieved 6 September 2016 from www.iea.org/topics/transport/.

IEA (2015) *CO_2 Emissions from Fuel Combustion Highlights (2015 Edition)*.

IEA (2016) *Mobility Model*. April 2016 version (database and simulation model), Retrieved 6 September 2016 from www.iea.org/etp/etpmodel/transport.

InterVISTAS (2015) *Economic Impact of European Airports – A Critical Catalyst to Economic Growth*. Retrieved 10 July 2016 from www.intervistas.com/downloads/ reports/Economic%20Impact%20of%20European%20Airports%20-%20January%20 2015.pdf.

Kuhnimhof, T., Armoogum, J., Buehler, R., Dargay, J., Denstadli, J.M. and Yamamoto, T. (2012) Men Shape a Downward Trend in Car Use among Young Adults – Evidence from Six Industrialised Countries. *Transport Reviews* 32(6): 761–779.

Kuhnimhof, T., Zumkeller, D. and Chlond, B. (2013) Who Made Peak Car, and How?: A Breakdown of Trends over Four Decades in Four Countries. *Transport Reviews* 33(3): 325–342. Retrieved 29 July 2016 from http://doi.org/10.1080/01441647.2013. 801928.

Lambert, F. (2017, 2 October) GM Announces Serious Electric Car Plan: 2 New EVs Within 18 Months, 20 Within 5 Years. *Electrek*. Retrieved 12 November 2017 from https://electrek.co/2017/10/02/gm-electric-car-commitment-new-models/.

Laurino, A., Ramella, F. and Beria, P. (2015) The Economic Regulation of Railway Networks: A Worldwide Survey. *Transportation Research Part A* 77: 202–212.

Levinson, D. (2015) Climbing Mount Next: The Effects of Autonomous Vehicles on Society. *Minnesota Journal of Law, Science and Technology* 16: 787.

Metz, D. (2013) Peak Car and Beyond: The Fourth Era of Travel. *Transport Reviews* 33(3): 255–270. Retrieved 29 July 2016 from http://doi.org/10.1080/01441647.2013. 800615.

Millard-Ball, A. and Schipper, L. (2011) Are We Reaching Peak Travel?: Trends in Passenger Transport in Eight Industrialized Countries. *Transport Reviews* 31(3): 357–378. Retrieved 29 July 2016 from http://doi.org/10.1080/01441647.2010.518291.

Mulligan, J. (2016, 14 April) Ryanair will Trial Transfer Traffic from this Summer. *Irish Independent*. Retrieved 10 July 2016 from www.independent.ie/business/irish/ryanair-will-trial-transfer-traffic-from-this-summer-34624914.html.

Nealer, R., Reichmuth, D. and Anair, D. (2015) *Cleaner Cars from Cradle to Grave: How Electric Cars Beat Gasoline Cars on Lifetime Global Warming Emissions*. Union of Concerned Scientists. Retrieved 29 July 2016 from www.ucsusa.org/sites/default/files/attach/2015/11/Cleaner-Cars-from-Cradle-to-Grave-full-report.pdf.

OECD/IEA (International Energy Agency) (2016) *Global EV Outlook 2016 – Beyond One Million Electric Cars*.

OECD/IEA (International Energy Agency) (2017) *Global EV Outlook 2017 – Two Million and Counting*. Retrieved 12 November 2017 from www.iea.org/publications/freepublications/publication/GlobalEVOutlook2017.pdf.

OICA (n.d.) Retrieved 6 September 2016 from www.oica.net/category/vehicles-in-use/.

Picone, M., Busanelli, S., Amoretti, M., Zanichelli, F. and Ferrari, G. (2015) *Advanced Technologies for Intelligent Transportation Systems*. Springer International Publishing. Retrieved 29 July 2016 from http://link.springer.com/chapter/10.1007/978-3-319-10668-7_1.

Polikarpov, A. (2015) 'Russian Railfreight Market at a Crossroads. *International Railway Journal* 55(9): 24–27.

Rayle L., Shaheen S., Chan N., Dai D. and Cervero R. (2014) App-Based, On-Demand Ride Services: Comparing Taxi and Ridesourcing Trips and User Characteristics in San Francisco. University of California, Berkeley Transportation Centre (UCTC), Working Paper.

Shaw, S., Fang, Z., Lu, S. and Tao, R. (2014) Impacts of High Speed Rail on Railroad Network Accessibility in China. *Journal of Transport Geography* 40: 112–122.

Small, K.A. and Verhoef, E.T. (2007). *The Economics of Urban Transportation*. London: Routledge.

Smith, G. (2017, 12 October) Paris wants to Ban the Combustion Engine by 2030. *Fortune*. Retrieved 12 November 2017 from http://fortune.com/2017/10/12/paris-combustion-engine-ban/.

Tran, M. (2013) Future Energy Mix and Transport, in M. Givoni and D. Banister (eds), *Moving Towards Low Carbon Mobility*. Cheltenham: Edward Elgar Publishing.

Tesla (n.d.) The World's Fastest Charging Station. Retrieved 12 November 2017 from www.tesla.com/supercharger.

Teslamotors (n.d.) Retrieved 12 November 2017 from www.tesla.com/supercharger.

UIC (n.d.) Retrieved 11 August 2016 from www.uic.org/highspeed.

UNCTAD Secretariat (2013) *Recent Developments and Trends in International Maritime Transport Affecting Trade of Developing Countries*. United Nations Conference on Trade and Development.

UNCTAD (2015) *Review of Maritime Transport*. United Nations Conference on Trade and Development. Retrieved 29 July 2016 from http://unctad.org/en/PublicationsLibrary/rmt2015_en.pdf

United Nations (2015a) *World Urbanization Prospects: The 2014 Revision* (No. ST/ESA/SER.A/366). New York: United Nations, Department of Economic and Social Affairs, Population Division. Retrieved 29 July 2016 from https://esa.un.org/unpd/wup/Publications/Files/WUP2014-Report.pdf.

United Nations (2015b) *World Population Ageing 2015* (No. ST/ESA/SER.A/390). New York: United Nations, Department of Economic and Social Affairs, Population Division. Retrieved 29 July 2016 from www.un.org/en/development/desa/population/publications/pdf/ageing/WPA2015_Report.pdf.

van der Laan, L. (1996) Changing Urban Systems: An Empirical Analysis at Two Spatial Levels. *Regional Studies* 32(3): 235–247.

Wang, Y., Teter, J. and Sperling, D. (2011) China's Soaring Vehicle Population: Even Greater than Forecasted? *Energy Policy* 39(6): 3296–3306.

Woldeamanuel, M.G. (2016) *Concepts in Urban Transportation Planning: The Quest for Mobility, Sustainability and Quality of Life.* Jefferson, NC: McFarland.

Wardsauto (n.d.) Retrieved 6 September 2016 from http://wardsauto.com/news-analysis/world-vehicle-population-tops-1-billion-units.

WHO (2016) *WHO Global Urban Ambient Air Pollution Database.* Retrieved 11 August 2016 from www.who.int/phe/health_topics/outdoorair/databases/cities/en/.

Wu, J., Nash, C. and Wang, D. (2014) Is High Speed Rail an Appropriate Solution to China's Rail Capacity Problems?. *Journal of Transport Geography* 40: 100–111.

Zhang, W., Guhathakurta, S., Fang, J. and Zhang, G. (2015) Exploring the Impact of Shared Autonomous Vehicles on Urban Parking Demand: An Agent-Based Simulation Approach. *Sustainable Cities and Society* 19: 34–45. Retrieved 29 July 2016 from http://doi.org/10.1016/j.scs.2015.07.006.

6 Geographies of religions

The religious factor in contemporary world politics – analytical frameworks

Anna M. Solarz

Let us briefly consider what it is that the following issues have in common: the establishment of the Islamic Republic of Iran and, more recently, the self-proclaimed caliphate known as the Islamic State; turmoil among Muslim fundamentalists in different parts of the world, including religiously motivated terrorism, especially after 9/11; internecine rivalry between Islamic factions in the Middle East and the policies of Islamist regimes in countries such as Turkey and Saudi Arabia; the influence and importance of the religion-based Hindutva movement in Indian politics, especially in relation to the Indian People's Party (BJP); the significance of different religions, as well as religious and anti-religious ethical systems in the politics of the United States, China, Latin America and black Africa; the influence of Judaism on Israel's internal and foreign policies, and the role of Buddhism in the countries of South-East Asia; the relationship between the Orthodox Church and the Kremlin in Putin's Russia; the role of religion in the formation of state identity and in the activities of international organizations; the influence of religion on transatlantic relations; the significance of the churches and Christian values for the emergence of the European Community; the ending of the Cold War and the overthrow of authoritarian systems in various parts of the world; the peaceful aims of *faith-based diplomacy* and interfaith dialogue; and the destructive role of religion in armed conflicts, especially those which are internal. All of the above-mentioned issues are examples of the impact of the religious factor on world politics. Many of these are analysed in the literature on international relations (see, e.g. Appleby 2000; Burgoński and Gierycz 2014; Curanović 2010; Johnston 2003; Johnston and Sampson 1994; Fox and Sandler 2004, 2006; Haynes 2006, 2007; Hoover and Johnston 2012; Jurgensmeyer 1994, 2003, 2008; Shah, Stepan and Toft 2012; Thomas 2005; Toft, Philpott and Shah 2011; Transatlantic Academy 2015). Many more are still awaiting description and analysis.

The influence of religion on contemporary world politics has been recognized and studied from as early as the 1990s (rarely before), but the vast majority of research on this subject has only appeared since 11 September 2001. It is estimated that the number of books on this subject published in the twenty-first century is six times higher than in last decades of the twentieth century (Hassner 2011: 37–56). The topic is still far from exhausted, though the focus is increasingly on systematizing the research that has already been carried out (Haynes 2007; Hoover and Johnston 2012; Shah *et al.* 2012; Toft *et al.* 2011). In addition, attempts are being made to incorporate the religious factor into international relations theory (Sandal and Fox 2013; Snyder 2011). However, religion remains one of the most important and at the same time, least understood and investigated security challenges in the twenty-first century (Warner and Walker 2011: 113). It is worth emphasizing that in the literature, but also in the press, media and daily conversations on the subject, there is a clear tendency towards a one-sided conceptualization of the role of religion as a factor which either positively or negatively affects people's lives, including international relations. The second of these approaches continues to dominate in the study of international relations, which is heavily influenced by rationalist theories (Kulska 2013: 99). Samuel Huntington (Huntington 1993, 1996) argued that the post-Cold War era was marked by the dysfunctional influence of religious divisions, but in the widespread criticism of this proposition, there was a surprising lack of discussion of the long-marginalized role of religion in international relations (Fox and Sandler 2004: 15). *Religion, the Missing Dimension of Statecraft* (Johnston and Sampson 1994) was published in answer to Huntington's dysfunctional vision of religion. Its authors attempt to show the lesser known and more positive face of faith as a factor which can facilitate reconciliation, peace and international cooperation (Kulska 2013: 95).

The complicated international reality of the role of religion is best described by the phrase 'the ambivalence of the sacred' (Appleby 2000). It depends on many factors (Philpott 2007), including the particular characteristics of a given religion, such as its universal aspirations and political theology, of which one aspect is the nature of the relationship it accepts with the political authorities. A religion's significance in international relations relates to the strength of its influence, the number of its adherents' – and demographic trends in this respect – internal divisions and geographical coverage. The history of the religion also seems to be important, as is the power of the tradition based on it, together with the successes and failures of the civilization connected with it. The vitality of a faith is reflected in the religiosity of its followers, but this is difficult to measure, and it is more difficult to determine its actual impact on the daily choices people make, including in the political sphere. Empirical research confirms, however, that in the twenty-first century, religion remains a significant factor in the lives of individuals and communities, and thus is an important component of 'soft power' in world politics (Haynes 2007). Many observers consider that religious adherence is the expression of a growing demand for a greater share in political life by those who have been marginalized and whose needs have not been satisfied by the secular state (Snyder 2011: 2–3). Others describe it as an expression of the rebellion of groups, nations and civilizations, which constitute the overwhelming majority of the Earth's population who are seeking their identity in a globalized world controlled by the West (Huntington 1996).

It is also worth noting, at the outset, a fundamental problem of research into religion: in the literature on religious studies, the conviction predominates that it

Box 6.1 Religion and politics in the United States

'E Pluribus Unum' ('from many, one') – the motto which appears on the Great Seal of the United States – goes to the very heart of religion in America: it is an affirmation of simultaneous unity and plurality, which seems paradoxical. However, many believe that America's hospitality towards people of different nationalities, cultures and religions is the real source of its strength. In 1967, Robert Bellah published an influential article titled 'Civil Religion in America', in which he defined 'the American way of life' as including a special religious coexistence and the collective faith that America serves a transcendent purpose in history as the prime agent of God (Bellah 1967). Historically, this has been closely aligned with various Christian denominations and Judaism. Especially since the 9/11 attacks, security grounded in a sense of unity has been called into question. Will the American paradox and existential motto survive unchanged?

The First Amendment of the United States Constitution states that 'Congress shall make no law respecting an establishment of religion or prohibiting the free exercise thereof', which means the confinement of politics and religion to separate realms. There is no tradition in America of political parties with a religious focus like the European Christian Democratic parties, but, as many have stated, 'religion has always played an important part in American politics'. Religious values and rhetoric were important in forming the new nation and religious ethics shaped the controversies over slavery and civil rights in the nineteenth and twentieth centuries. More than 70 per cent of American adults still profess Christianity (in 2007, 78.4 per cent) and the USA remains home to more Christians than any other country in the world. However, the share of Americans who are religiously unaffiliated or identify with non-Christian faiths has risen (to 22.8 and 5.9 per cent respectively, from 16.1 and 4.7 per cent in 2007). Americans are among the most religious of Western nations, which contradicts secularization theory, as the USA is arguably the most modern country in the world. According to Pew Research Center data, 63 per cent of Americans are absolutely certain of their belief in God, and 20 per cent are fairly certain. For 53 per cent of American adults, religion in their life is very important and for 24 per cent it is somewhat important. Only 11 per cent treat religion as something not at all important. The most religious people are to be found among the Jehovah's Witnesses, historically black Protestants, Mormons and Evangelical Protestants (Pew Research Center 2015c).

The religiosity of Americans has an impact on the country's political landscape. It is said that Americans could elect almost anyone except an atheist. Even among supporters of the more secularist Democratic Party, the majority want their candidate to be religious. Since the 1970s, what is known as the Religious Right, made up of Protestant fundamentalists and conservative evangelicals, has become increasingly influential in domestic policy, and from the 1980s onwards in foreign policy as well. The events of 9/11 and the subsequent 'war on terror' were opportunities for President G.W. Bush and his neoconservative advisers and political allies to introduce their ideas into US foreign policy. The wars in Afghanistan and Iraq were the outcome. In 2006, LaFranchi identified elements in US foreign policy which he called 'evangelized foreign policy', such as: The International Religious Freedom Act (1998) which makes freedom of religion and conscience a 'core object' of US foreign policy; The Trafficking Victims Protection Act (2000), against crime syndicates engaged in human trafficking; The Sudan Peace Act (2002), against attacks on Christians and animists in the south of the country, which led in consequence to the independence of Southern Sudan. In addition, conservative Christians were involved in the Bush administration's focus on AIDS in Africa (which helped to stop the epidemic there) and in the activities of anti-family planning organizations worldwide (Haynes 2007: 255). The religious sens-itivity of the political decision makers of the world's greatest power seems an unmistakable factor in its foreign policy. The significance of the religious factor in this context is also recognized by Madeleine Albright, Secretary of State during Bill Clinton's presidency, in her book *The Mighty and the Almighty: Reflections on America, God, and World Affairs* (New York 2006).

is impossible to define this phenomenon in a way which encompasses the world's religious diversity (Bronk 2009). The word 'religion' is of Latin origin and can express, on the one hand, the worship of God/gods (the external aspect of religiosity, which we can perceive, study and to some extent measure: doc-trine, ethics, forms of worship and related religious institutions). On the other hand, religion is the mutual bond between humanity and the Absolute/Divine (the inner aspect, real only to the believer, invisible to the eye and eluding empirical investigation). This second aspect is the essence of religion, whose real impact on human life is very difficult to determine because of its unique and completely 'unscientific' character. By limiting our research to doctrine, ethics, forms of worship and religious organization, we are usually unable to grasp the essence – that is, what it is which really motivates believers to behave in a spe-cific way and make specific choices and decisions, including in the sphere of politics. In today's world, however, we cannot afford to eliminate religion from our field of research, because doing so would completely distort the reality we observe. However, we cannot now, and may never be able to, scientifically explain religion. In light of this, Galileo's conviction that the task of science is to explain how the heavens turn and not how to get there (which falls to religion) has not lost its relevance in the twenty-first century. Therefore, the social sci-ences, given their interest in explaining human behaviour, should make a priority of incorporating the study of the religious factor. International relations is no exception in this respect.

Religion and international relations: new research perspectives

In the twentieth century, our conceptions concerning the time-and-space reality surrounding us changed dramatically under the influence of new discoveries in the field of physics. The construction of the atomic bomb and new methods for the transmission of information revolutionized the world. This, in turn, was reflected inevitably in philosophy, and thence in the social sciences, in which, for example, attempts were made to apply quantum theory as a way of explain-ing human behaviour (Wendt 2015). It was felt that an opportunity had presented itself to move towards the unification of physical and social ontology, one of the most desired goals of metascience. It seems that, indirectly, this was also behind a surge of interest in the influence of cultural factors, including religious, on international reality. In many cases, the 'cultural' approach radically alters the way we explain this reality, although so far this has not led to the emergence of a new paradigm in the Kuhnian sense. Nevertheless, there is a growing awareness that religion is a factor which so strongly affects human society that its absence from our analysis of social, including international, reality, leads us to incomplete and untrue conclusions.

In varying degrees, religion undergirds the choices of individuals, families, groups and entire communities within countries, thereby permeating the formulation governmental policies and the international activities of other actors. It is a component of ethnic and national identity, and therefore contributes to the definition of foreign policy interests and objectives (Telhami and Barnett 2002). We can also observe the influence of religion at the interstate level, especially regional and systemic (global), facilitated by the processes of globalization (Thomas 2005: 28). At every level, religion exercises its influence in two ways: by what it 'says' in its doctrine and theology, and by what it 'does' as a 'social phenomenon and mark of identity' (Haynes 2007: 12). But its impact and importance is very difficult to measure. This is one of the reasons why many international relations specialists exclude the religious factor or treat it as marginal. It must be stressed, however, that, in the last decades, the techniques of sociological research have improved considerably, and therefore also the quality and credibility of data collected, including concerning religion and religiosity, and this should be reflected in the study of international relations (cf. Wendt 2008: 14).

Another reason for the neglect of religion in the social sciences is the deeply rooted paradigm of secularization, which posits the inevitable decline of the importance of religion in social life in parallel with modernization and growth in wealth. International relations as the most Western-centric of the social sciences is still dominated by this way of thinking, although in related disciplines – especially sociology – it has already been discredited (Fox and Sandler 2004: 16). The secularization and privatization of religion, which we have seen in the societies of Western Europe, contrasts with the growing importance of the religious factor in different regions outside the Old Continent. Peter Berger, whose book *The Sacred Canopy* was published in 1967 (cf. Berger 1967) contributed to the consolidation of the paradigm of secularization in sociology, subsequently described – in a work published thirty-two years later under his editorship 'The Desecularization of the World' – his 'aha! experience' when he understood while studying religious fundamentalism that 'what is rare is not the phenomenon itself [i.e. religion – A.M.S.] but knowledge of it'. He went on to emphasize the fact that the contemporary world with some exceptions is 'furiously religious', which means that the entire literature (including his own) on the theory of secularization is erroneous (Berger 1999: 2).

Research into religion therefore needs a new framework and a new research perspective, which can be derived from scientific advancements and in-depth social studies of the phenomenon of religion in the world around us. The assumptions of the natural disappearance of religion in parallel with moderniza-tion, and of science as an alternative, and 'better', way of explaining the world, on which, in part, the dominant, rationalist theories of international relations were based, have proven to be wrong. In contrast to the way the reality surround-ing us has hitherto been presented, the principle of quantum mechanics opens up new fields, which seem to make room for a 'divine hypothesis' as the key to the mystery of existence (Horodeński 2012). And while it is unlikely that science will ever prove the existence of God or gods, it also does not exclude his/their existence, as is increasingly reflected in philosophy and the social sciences which are ontologically ever closer to the natural sciences.

The world's most important religions in numbers

Religion is a phenomenon which is permanently intertwined with human history. According to Erich Fromm, there is no culture of the past, and it seems there can be no culture in the future, which does not have religion in broad sense (Fromm 1950: 21).[1] The literature estimates the number of different religions in the world at about 10,000, but only a dozen or so are, for various reasons, considered to be among the most important (Johnson and Grim 2013: 12). These are usually: Christianity, Islam, Hinduism, Buddhism, Chinese Folk-Religion, Judaism and Ethnoreligions (Fromm 1950: 21), or: Christianity, Islam, Hinduism, Buddhism, Confucianism, Yoruba religion, Judaism and Daoism (Prothero 2010; Juergens-meyer 2006); On the Adherents.com website, the ten main religions are listed as: Christianity, Islam, Hinduism, Buddhism, Sikhism, Judaism, Baha'ism, Confu-cianism, Jainism and Shintoism (The Christian Science Monitor n.d.). If we add to this last list Ethnoreligions (including Yoruba religion) and Daoism/Taoism, we have the world's basic twelve religions, whose adherents together account for about 87 per cent of world population (Johnson and Grim 2013; Adherents 2005). Sometimes, the Zoroastrians (known in India as Parsis, i.e. 'Persians') are included in this group, in contrast to a variety of tribal beliefs which, unlike Zoroastrianism, do not form a single religious system (Adherents 2005).

It is worth noting that the demography of religion also includes the category of people who not associated with any religion (unaffiliated). In 2010, this was 16.4 per cent of the world's population (1.1 billion) (Hackett, Stanowski, Potančoková, Grim and Skirbekk 2015). Around half of this group are agnostics[2] and atheists,[3] and the rest are not associated with any particular religion but declare themselves to be theists or believers in 'something' supernatural (Adher-ents 2005). Studies covering 99.98 per cent of the world's population in 198 countries show that if current trends continue up to 2050, the unaffiliated cat-egory will shrink to 13.2 per cent or, with different variables, 14.3 per cent of the planet's inhabitants. The number of unaffiliated will grow in North America, Europe and Latin America, while in all other parts of the world their numbers will fall (Hackett *et al.* 2015). It is important to emphasize that in 2010, 62 per cent of the entire unaffiliated group, or around 700 million, lived in China.[4] The

majority of the Czech, Estonian, Hong Kong, Japanese and North Korean populations were also unaffiliated in 2010 (Hackett *et al.* 2015: 831).[5] However,
demographic data shows that in all of these countries the unaffiliated are mainly
older people. Minority, though growing, unaffiliated groups consisting mainly of
young people are found in the USA, Australia, New Zealand and some countries
in Europe (Hackett *et al.* 2015).

Data presented by The World Religion Database (WRD) differs only slightly
from the above, which is based primarily on research by the Pew Research
Center (PRC). The WRD treats agnostics and atheists as separate groups,
accounting for 9.8 per cent and 2.0 per cent of the world population respectively
in 2010 (so a total of 11.8 per cent are non-religious according to these figures).
In 1910, both groups accounted for just a fraction of a per cent (0.2) in total.
After the Bolshevik revolution and the establishment of the communist system
in many countries of the world, this figure reached 6.7 per cent in the mid-
twentieth century and peaked in 1970 at 19.2 per cent (14.7 per cent – agnostics,
4.5 per cent – atheists). In 2000, these groups shrank to 12.9 per cent of world
population (Johnson and Grim 2013: 42); and a further decrease in the proportion of non-religious people in the world is also expected in the light of WRD
projections.

According to WRD data, in 2010, out of a world population of 6.9 billion, there
were about 2.26 billion Christians (32.8 per cent of the total), 1.55 billion
Muslims (22.5 per cent), 0.95 billion (13.8 per cent) Hindus, some 0.49 billion
(7.2 per cent) Buddhists, 0.44 billion (6.3 per cent) Chinese Folk-Religionists,
0.24 billion (3.5 per cent) Ethnoreligionists, 63 million (0.9 per cent) New Religionists, almost 24 million (0.3 per cent) Sikhs, 14.7 million (0.2 per cent) Jews,
13.7 million (0.2 per cent) followers of spiritism, 8.4 million (0.1 per cent)
Daoists/Taoists, 7.3 million (0.1 per cent) Baha'is, 6.4 million (0.1 per cent)
Confucianists, 5.3 million (0.1 per cent) Jains and a fraction of a per cent of
Shintoists (about 2.7 million) and Zoroastrians (197,000) (Johnson and Grim
2013: 10).

PRC data included in the latest edition of *The Changing Global Religious
Landscape* (as of 2015) is somewhat different from the WRD data for 2010.
According to the former, Christians number 2.27 billion (or 31.2 per cent of the
world total), Muslims – 1.75 billion (24.1 per cent), Hindus – 1.39 billion (15.1
per cent) and Buddhists – 500 million (6.9 per cent). Traditional religionists
(including the Chinese varieties of these beliefs, Native American religions and
Australian aboriginal religions) totalled some 418 million (5.7 per cent). Approximately 59.7 million (0.8 per cent of the world's population) were adherents of
Sikhism, the Baha'i Faith, Jainism, Daoism/Taoism, Tenrikyo, Wicca and Zoro-
astrism (in the remainder of the chapter, all these religions are referred to using
the PRC category for them of *other religions*). Judaism has some 14.2 million
adherents (0.2 per cent of the world's population) (Pew Research Center
2017: 10).

Among Christians,[6] the largest religious group in the world, in 2010, according to WRD, 51.5 per cent were Catholics, which represented 16.9 per cent of

Box 6.2 The Roman Catholic Church/Holy See in world politics

The Roman Catholic Church/Holy See is a transnational religious actor and, at the same time, an entity in international law. The right of the Pope – as the head of the Catholic Church and leader of the Church State (*Patrimonium Sancti Petri*) founded in the sixth century – to engage in international relations has been recognized since the Middle Ages. The liquidation of the Church State in 1870 brought about no changes in this regard. In addition, on 11 February 1929, the Lateran Treaty was signed between the Holy See and Italy, which created the sovereign State of Vatican City as the territorial guarantee of papal authority (with an area of 44 hectares, it is the smallest of the world's states). Today, the Catholic Church is the only entity in international relations, which combines the characteristics of a transnational religious actor and a *sui generis* entity in international law with all the privileges due to states (including the right of active and passive legation, the right to conclude international agreements and participate in the work of intergovernmental organizations). As the Holy See or the Vatican, the Catholic Church also acts in the international arena, for example, by maintaining diplomatic relations with more than 180 states, as well as with the EU and the Order of Malta, and working with many international organizations as a member or observer. The Holy See has bilateral agreements with other states, known as concordats, whose purpose is primarily to regulate the status of the Catholic Church in a given country. Some concordats are of a special character, such as the concordat with Morocco, which was established in the form of an exchange of letters between King Hassan II and Pope John Paul II in 1984, and the Fundamental Agreement signed with Israel on 30 December 1993. The Holy See is also a party to several international conventions.

In the international arena, the representatives of the Holy See seek to draw attention to the ethical dimension of this sphere of social relationships (especially human rights), drawing on the 125-year tradition of Catholic social teaching, begun by Leo XIII's encyclical Rerum Novarum Encyclical 'On Human Labour'. Important papal contributions to international relations in the post-Second World War period include: the encyclicals Pacem in terris (1963), Populorum progressio (1967), Solicitudo rei socialis (1987), Centesimus annus (1991) and Laudato si (2015), annual messages on the World Day of Peace (celebrated by the Catholic Church on 1 January since 1968), as well as speeches at gatherings of international institutions.

The Holy See's intention with regard to international relations is to introduce an ethical dimension (it describes itself as 'the conscience of humanity' and 'an expert in humanity'). The most important aims of its activity are the protection of human rights, including those of communities such as the family or the nation, the maintenance of peace and reconciliation between nations, and the integral development of the 'human family'. Examples of this include John XXIII's contribution to alleviating the Cuban crisis in 1962, and John Paul II's part in bringing about the end of the Cold War confrontation and the fall of many dictatorships. The latter also opposed the wars with Iraq in 1991 and 2003, but invoked the right of humanitarian intervention during the conflict in the former Yugoslavia. Pope Francis was among those who were instrumental in restoring US–Cuban relations during Barack Obama's presidency. The Holy See and Catholic organizations have frequently been mediators in domestic and international disputes, especially in Latin America and Africa. Since the 1960s, especially within the United Nations, the Holy See has been working to promote a comprehensive concept of development (described in the Church's teaching as 'integral and solidarity humanism'), which will encompass 'all people' without exclusion (in terms of international relations, this means all countries, including the poorest) and the 'whole person', both the material and spiritual dimension. In the 1990s, in the face of accelerated globalization, the Holy See became involved in the work of the UN system to an unprecedented degree, which included playing a significant role in the preparation of UN 'global conferences' (e.g. Rio 1992, Vienna 1993, Cairo 1994, Copenhagen 1995, Beijing 1995 and Durban 2001). However, the Church has opposed the propagation within the UN of a new paradigm of health under the umbrella of which can be found reproductive rights, including artificial contraception and the 'right' to abortion, as well as the redefinition of the traditional family and its rights. The Church's stance has in turn met with criticism from those who wish to make globalization a transmission belt for cultural change in the world.

total world population (in 1910 this was 47.6 per cent and 16.6 per cent respectively). In 2010, 18.6 per cent of all Christians were Protestants, representing 6.1 per cent of world population (18.8 per cent and 6.5 per cent respectively in 1910). The Orthodox Church was the branch of Christianity which suffered the most due to the communist system: in 2010, its followers constituted 12.2 per cent of all Christians and 4 per cent of the world population (100 years earlier this was 20.4 per cent and 7.1 per cent respectively). Over the 100-year period, the Anglican Church had also seen large numerical change – in 2010, it represented 3.8 per cent of all Christians and 1.3 per cent of world population (down from 5.4 per cent and 1.9 per cent in 1910). Independent churches, especially in Africa and Asia, have seen significant increases over the 100 years: from 1.5 per cent of all Christians in 1910 to 15 per cent in 2010 (their share of world population increased during the period from 0.5 to 4.9 per cent). Groups on the fringes of mainstream Christianity (Marginals)[7] also grew in the period, from 0.2 per cent of all Christians in 1910 to 1.5 per cent (from 0.1 to 0.5 per cent of world population) (Johnson and Grim 2013: 14–16). The countries with the largest populations (in absolute numbers) of Christians in 2010 were: the United States (nearly 248 million), Brazil (177.3 million), Russia (116 million), Mexico (nearly 109 million), China (106 million), the Philippines (nearly 85 million), Nigeria (73.6 million), DR Congo (62.6 million), Germany (57.6 million) and India (57 million) (Johnson and Grim 2013: 15).

Followers of the religion begun by the revelations received by the Prophet Muhammad in the seventh century AD constituted 12.6 per cent of world population in 2010. In 2010, this had risen to 22.5 per cent (Johnson and Grim 2013: 19). The proportions of Sunnis and Shiites in the total Muslim population have not changed – in 1910 there were 86.2 and 12.9 per cent respectively, and in 2010, 85.6 and 12.9 per cent. In the same period, there was an increase in the

Box 6.3 Political Islam in world politics

Changes in the policy of Muslim governments in the direction of actively promoting Islamic law and morality in their countries and abroad were already being demanded by Islamic intellectuals and activists in the second half of the nineteenth century (the An-Nahda renewal movement). These demands grew in intensity in the first half of the twentieth century (advocated by such figures as Hassan al-Banna, the founder of the Muslim Brotherhood, and the Islamic ideologist Sayyid Qutb). By the end of the twentieth century, they had been implemented by only a few countries, but had increasing support in Muslim society. Despite many regional and local differences, religious, ethnic and clan divisions, a multitude of languages and dialects, the followers of Allah have a feeling of belonging to one Muslim community – the *Ummah*. In 1924, the symbol of this unity, the title and office of the Caliph, was abolished by decision of the Kemalist parliament in Turkey. In the late 1960s and early 1970s, the idea of unity was reborn in the form of the Organization of the Islamic Conference organization, which gradually expanded and in 2011, was renamed the Organization of Islamic Cooperation. It now comprises fifty-seven countries on four continents and is the only governmental organization in the world in which membership is based on religion. Demographically strong or affluent Muslim states compete for leadership in this group (primarily Saudi Arabia, Iran and Turkey and, to a lesser extent, Egypt and Pakistan). During the Cold War, the United States used Islam to fight the USSR by providing support and weapons to religiously motivated groups. This was especially evident during the Soviet intervention in Afghanistan between 1979 and 1989. In the 1990s, many of these combat-experienced Islamic militants turned against their own 'insufficiently religious' governments in the Middle East and Africa. The West was determined not to allow the Islamists to take over power, and so supported the 'secular regimes', as a result of which the 1990s were marked by bloody conflicts such as the civil war in Algeria. Islamic fundamentalism also turned its attention to the USA, which, after the ending of the Cold War, had became more active in the Arab world, due, in part, to the 1991 war in Kuwait and the Oslo Accords of 1993. Tangible evidence of this growing involvement was the increase in the US military presence, the intervention in Somalia and support given to friendly Arab states, especially the despotic Saudi Arabia and authoritarian Egypt. The struggle against the 'demoralized' West, whose political, economic and cultural embodiment was increasingly seen as the United States, was taken up in the name of religion by non-governmental actors who resorted to terrorism on an ever-greater scale as a means of war against overwhelmingly superior governmental forces ('asymmetric conflicts'). These groups also gained a propaganda advantage from the United States' support for Israel with its aggressive policy of retaliation against pro-Palestine activists, some of whom also used terrorist methods. The Al-Qaeda attacks in the USA proved effective, inspiring hope that the collapse of the demoralized West was near at hand. The existing Islamic and Salafic terrorist groups were joined by new ones, encouraged by the unexpected success of this method of struggle. Their targets expanded to include, among others, the countries of Western Europe where they could take advantage of the growing European part of the *Ummah*. What followed were numerous, large and shocking terrorist attacks in Europe, the USA and elsewhere.

proportion of groups on the fringes of Islam (especially small communities living in the Middle East and South Asia: Khawarij, Ahmadiyya, Alawites, Druze and Yazidis): from 0.5 to 1.3 per cent of all Muslims. It is worth noting the decline in the proportion of people identifying with Sufism (not treated as a separate group within Islam) – from 39.9 per cent of Muslims in 1910 to 19.9 per cent in 2010. Islam, thus, seems less mystical today than previously. Although Islam was born in the area of present-day Saudi Arabia and is firmly associated with the Arabs, the Middle East and North Africa, its centre is increasingly moving eastwards. In 2010, countries with the largest populations of Muslims (home to 65.8 per cent of all Muslims) were: Indonesia (209.1 million, i.e. 13.1 per cent of all Muslims), India (176.2 million and 11 per cent), Pakistan (167.4 million and 10.5 per cent), Bangladesh (134.4 million and 8.4 per cent), Nigeria (77.3 million and 4.8 per cent), Egypt (76.9 million and 4.8 per cent), Iran (73.5 million and 4.6 per cent), Turkey (71.5 million and 4.5 per cent), Algeria (almost 35 million and 2.2 per cent), Morocco (31.8 million and 2 per cent) (Pew Research Center 2015a).[8]

In addition to Christians and Muslims, Hindus account for a large proportion of the world's population (13.8 per cent in 2010, compared to 12.7 per cent in 1910), mainly in South Asia: India (893.6 million), but also Nepal (20.2 million), Bangladesh (14 million), Indonesia (3.8 million), Sri Lanka (2.7 million), Pakistan (2.2 million), Malaysia (1.7 million), the United States (1.4 million), South Africa (1.1 million) and Myanmar (0.8 million) (Johnson and Grim 2013: 25). Hinduism is one of the world's oldest surviving religions. It has a common canon of sacred writings but no clear leader, and is made up of many closely related traditions, which range in character from monotheism to polytheism. Most Hindus are Vaishnavas (adherents of Vishnu, representing 9.3 per cent of the world's population and 67.6 per cent of all Hindus), others are Shaivas (Shiva adherents, 3.7 and 26.6 per cent respectively), Shaktas (Shakti adherents, 0.4 and 3.2 per cent) and small groups who follow Neo-Hinduism (0.3 and 2.1 per cent) and Reform Hinduism (0.1 and 0.5 per cent) (Johnson and Grim 2013: 26).

Chinese Folk-Religionists accounted for 6.3 per cent of world population in 2010, according to WRD data (22.2 per cent in 1910). They live mainly in China and Taiwan and other places where the Chinese diaspora is located. Chinese Folk-Religionists represent 10.4 per cent of all Asians, 0.3 per cent of the inhabitants of Oceania, 0.2 per cent of the population of North America (most – approximately 670,000 – live in Canada) and 0.1 per cent of Europeans (Johnson and Grim 2013: 32). Chinese Folk-Religion is deeply intertwined with the history of China, and contains elements of animism, shamanism and magic, as well as the cult of ancestors and the veneration of countless deities. It is known primarily for its rituals and festivals such as the Chinese New Year. Many of its followers also practise other Chinese traditions – such as Buddhism, Confucianism and Daoism/Taoism without any feeling of discomfort. If we consider that China is simultaneously a country where 62 per cent of the world's religiously unaffiliated live, the expression 'unity in diversity' used by the Chinese philosopher Chung-Ying Cheng to describe his compatriots seems very appropriate (Johnson and Grim 2013: 31).

According to WRD data, the Buddhist Sangha (the community of those who follow the teachings of Buddha) accounts for 7.2 per cent of world population (7.9 per cent in 1910). It is a very diverse religion, with many different branches. It is active in various regions of the world, but the majority of Buddhists live in Asia (where it accounts for 11.7 per cent of the total population), and especially East and South-East Asia. In Oceania, 1.6 per cent of the population are Buddhists, 1.3 per cent in North America, 0.2 per cent in Europe and 0.1 per cent in Latin America (Johnson and Grim 2013: 34). In the twentieth century, Buddhism began to emerge from Asia – small groups appeared in many countries of Latin America and Europe. This was due in part to the missionary spirit of this universalist religion, but also to the fact that Asia had become an attractive travel destination. Not without significance were the personal qualities of the 14th Dalai Lama, leader of Chinese-occupied Tibet, who garnered much admiration in the West. Interestingly, in the twenty-first century, Buddhism is growing fastest in Qatar, the UAE and Mozambique (Johnson and Grim 2013: 38). In Asia, there are a large number of non-Buddhists, especially in China, whose daily way of life is strongly influenced by Buddhism due to its entrenchment in the region. Many such Chinese thus affected consider themselves to be unbelievers or followers of Chinese Folk-Religion. There are three main varieties of Buddhism. The most conservative and oldest variety or school is Theravada, which is particularly strong in South-East Asia. In 2010, 35.8 per cent of all Buddhists (2.6 per cent of the world's population) identified with this school. Adherents of the Mahayana school, according to which the *bodhisattva*[9] have a special role in communicating the teachings of Buddha to others, accounted for 53.2 per cent of all Buddhists (3.8 per cent of the world's population), and is strongest in China, Japan, Korea and Singapore. Last, the Tibetan school (Lamaism) and its variants, which other than in Tibet is also popular in Mongolia, accounted for 5.7 per cent of all Buddhists (0.4 per cent of the world's population) (Johnson and Grim 2013: 38).

Ethnoreligionists make up 3.5 per cent of the world's population (about 400 million adherents) according to WRD data. These are various groups which follow natural religion, and include animists, spirit-worshippers, shamanists, ancestor-venerators and polytheists, as well as local or tribal folk-religionists. Ethnic religions are usually based on oral tradition, such as myths, rituals and rites passed down from generation to generation (Johnson and Grim 2013: 38). In these religions, the sacred sphere mixes with the profane and permeates the daily lives of believers, who rarely consider the spirit world as separate from the real world. Under the influence of missionary religions, especially Islam and Christianity, the number of followers of the traditional religions is falling, but often, through a merging of the boundaries between old and new, syncretic religions are formed which are difficult to classify unambiguously. In 1910, ethnic religions were practised by 58 per cent of Africans, 5.6 per cent of Asians, 0.2 per cent of Europeans, 3.5 per cent of South Americans, 0.2 per cent of North

Americans and 19.7 per cent of Oceanians (including 83.3 per cent of Melanesians and 23.3 per cent of Micronesians, but interestingly only a small fraction of a per cent of Polynesians and 1.2 per cent of Australians and New Zealanders) (Johnson and Grim 2013: 39). In 2010, the figure for Africa had fallen to 8.7 per cent, in Asia it was 3.5 per cent, 0.2 per cent in Europe, 0.6 per cent in South America, 0.4 per cent in North American and 1 per cent in Oceania. The countries with the largest numbers of Ethnoreligionists today are: China (57.8 million, which is in addition to the number of Chinese Folk-Religionists), India (45.8 million) and Nigeria (12.1 million). Vietnam, Madagascar, South Korea, Mozambique, Indonesia, Tanzania and Ethiopia all have populations of between 5 and 10 million Ethnoreligionists (Johnson and Grim 2013: 39).

The demography of religion includes the category of New Religionists. These are members of various Hindu or Buddhist sects, and also syncretic religions, which combine elements of Christianity and Eastern religions (Johnson and Grim 2013: 41). In 2010, the number of New Religionists amounted to 63 million or about 0.9 per cent of the world's population, primarily in Asia (especially in Japan and Indonesia) and Latin America. The largest of the new religions in terms of the number of adherents is Soko Gakkai International (18.5 million adherents or 29.4 per cent of all New Religionists), which was founded in Japan in 1930 on the basis of Buddhism, and since 1975 has been an international movement (Johnson and Grim 2013: 45).

Smaller religions listed in the statistics are Sikhism (23.9 million) (Johnson and Grim 2013: 48), North Korea's cult of the Great Leader and related Juche ideology (19 million) (Adherents 2005), Spiritism (15 million) (Adherents 2005; Johnson and Grim 2013: 54), Judaism (14.7 million),[10] Daoism/Taoism (8.4 million) (Johnson and Grim 2013: 57), the Baha'i Faith (7.3 million), Confucianism (6.4 million),[11] Jainism (5.3 million), Shinto (2.7 million) (Johnson and Grim 2013: 69),[12] Caodaism (4 million) (Adherents 2005), Zoroastrianism (2.6 million) (Adherents 2005)[13] and Tenrikyo (2 million) (Adherents 2005). The same data indicates that there are about 1 million neopagans in the world and less than 1 million adherents each of Unitarian Universalism, Rastafarianism and Scientology (Adherents 2005).

The geographical distribution of individual religions is worth considering in light of the fact that the population of the Asia-Pacific region accounts for 58.8 per cent of the world's population, sub-Saharan Africa for 11.9 per cent, Europe for 10.8 per cent, Latin America for 8.6 per cent, North America for 5.0 per cent and the Middle East and North Africa for 4.9 per cent. According to the 2012 PRC report, 99 per cent of Hinduism is located in the Asia-Pacific region, as is Buddhism and about 90 per cent of the Ethnoreligionists (just some 10 per cent of the adherents of these religions live in sub-Saharan Africa and Latin America taken together). The vast majority (89 per cent) of the followers of *other religions* are also located in the Asia-Pacific region. This is also true of 62 per cent of Muslims (the remaining 20 per cent are in the Middle East and North Africa, 16 per cent in sub-Saharan Africa and the rest mainly in Europe). The vast majority of people who are not associated with any religion (unaffiliated) – 76 per cent – are also inhabitants of the Asia-Pacific region (nearly 700 million of whom are Chinese). The Christian religion is the most evenly distributed throughout the world: 26 per cent of Christians live in Europe; 24 per cent in Latin America and the Caribbean together; 24 per cent also in sub-Saharan Africa; 13.2 and 12.3 per cent in the Asia-Pacific region and North America respectively; and 0.6 per cent in the Middle East and North Africa. The Jews live in almost equal proportions in the Middle East (Israel) and North America, a lower number in Europe and small Jewish communities are found in all other regions of the world (Pew Research Center 2012: 10).

An analysis of the age of religious adherents also gives interest results. The average age worldwide is twenty-eight. The average ages of Muslims (twenty-three) and Hindus (twenty-six) are below this average, and all other religions are above it (Christians – thirty, other religions – thirty-two, tribal religions – thirty-three, Buddhists and unaffiliated – thirty-four, Jews – thirty-six). It seems therefore that we may expect especially the first two religions mentioned to have high birth rates in the near future resulting in an increase in the number of their adherents (Pew Research Center 2012: 13). The 2015 Pew Research Center survey confirms this with respect to Islam – if current trends persist, the number of Muslims in the mid-twenty-first century will be almost equal to that of Christians (2.76 billion versus 2.92 billion), which means that in the near future, Islam will see the largest numerical growth – in 2050, the percentage of Muslims in the world's population will have increased from the current 23.2 to 29.7 per cent. The number of Christians will remain at the current level (31.4 per cent). The proportion of Hindus will not increase (currently 15 per cent, in 2050 – 14.9 per cent), even though their absolute number will increase by more than 300 million (in 2050 – 1.38 billion). The proportion of Jews will remain at the current level (0.2 per cent), while that of adherents of other religions will decrease slightly (from 0.8 per cent to 0.7 per cent). The share of Buddhists in the world's population will drop significantly (from 7.1 per cent to 5.2 per cent) as will that of the adherents of Ethnoreligionists traditional religions (from 5.9 per cent to 4.8 per cent). In 2050, the proportion of religiously unaffiliated will also have decreased – from over 16 per cent to 13.2 per cent (Pew Research Center 2015b: 6).

The future of the world's religiosity and religion

Demographic change is one of the most important factors affecting international relations in the globalized world, as it leads to cultural change. It is worth noting that according to long-term projections presented in the PRC survey already mentioned, in 2070, the proportions of Christians and Muslims in the world will be equal (32.3 per cent), and in 2100, the latter will exceed the former by more than 1 percentage point (34.9 versus 33.8 per cent) (Pew Research Center 2015b: 14).

Although such long-term predictions should always be treated with caution, the above data shows that the future belongs to two universalist religions. Both are internally divided, but Islam seems almost uniform in comparison to the diversity of Christianity, which according to the World Christian Encyclopedia has at least 33,000 separate denominations (Barrett, Kurian and Johnson 2001:16).[14] Perhaps this heterogeneity facilitates the spread of Christianity in the face of the varied spiritual needs of people throughout the world. Nevertheless, one of the greatest challenges for the Christian religion in the twenty-first century is the building of unity, which would serve to preserve its common values in the encounter with other religions, especially Islam.

In his book *The Next Christendom: The Coming of Global Christianity*, Philip Jenkins carefully analyses the available data on the future of religion and religiosity and expresses the conviction that the shape of the coming world will be determined by biological and demographic factors, and that a good knowledge of the religious aspect is essential to understanding the emerging world order of the twenty-first century. He argues that without peace between religions there can be no question of peace between peoples. The impact of the religious factor on human behaviour fully justifies it being the subject of study within the discipline of international relations. Jenkins notes that traditional Christianity is fading in the global North, but is being strengthened and revived by the Churches of the South through immigration and evangelization. In his view, this is likely to result in Christianity developing a 'southern character', that is, being conservative and charismatic (Jenkins 2009: 313–314), which in turn would imply an increase in the importance of religion in social relations.

PRC research confirms that over the last 100 years, Christianity has become a non-European religion: in 1910, as many as 66.3 per cent of all Christians lived in Europe, 27.1 per cent lived in the Americas, 4.5 per cent lived in the Asia-Pacific region, 1.4 per cent lived in sub-Saharan Africa and 0.7 per cent lived in the Middle East and North Africa (MENA). In 2010, only 25.9 per cent of Christians now lived in Europe, while 36.8 per cent lived in the Americas, 23.6 per cent lived in sub-Saharan Africa, 13.1 per cent lived in the Asia-Pacific region and 0.6 per cent lived in the MENA region. The proportion of Christians in the world's population fell slightly – from 35 per cent in 1910 to 32 per cent in 2010. Countries with the largest Christian populations today are the United States (nearly 80 per cent of the population profess this religion, and account for 11.3 per cent of all Christians worldwide), Brazil (90.2 per cent of the population and 8 per cent of all Christians), Mexico (95 per cent of the population and 4.9 per cent of all Christians respectively), Russia (73.6 per cent of the population and 4.8 per cent of all Christians), The Philippines (93.1 per cent of the population and 4 per cent of all Christians), Nigeria (50.8 per cent of the population and 3.7 per cent of all Christians), China (5 per cent of the population and 3.1 per cent of all Christians), DR Congo (95.7 per cent of the population and 2.9 per cent of all Christians), Germany (70.8 per cent of the population and 2.7 per cent of all Christians) and Ethiopia (63.4 per cent of the population and 2.4 per cent of all Christians). It is notable that this list contains only one European country. Now 60.8 per cent of all Christians live in the global South and only 39.2 per cent in the global North, although in the North, Christianity is still the predominant religion (69 per cent consider themselves to be Christians), and in the South, Christians are in the minority (23.5 per cent of the total population). The great changes seen in this regard over the last 100 years form the basis of the arguments put forward by Jenkins (Pew Research Center 2011: 9).

However, Jenkins' book also draws the reader's attention to the fact that it is not only affiliation to particular religious communities which determines the influence of the religious factor on world politics. In addition to the activities of the religious organizations themselves, international relations are impacted by the activities undertaken by religiously motivated people, whose behaviour results from their faith and the religious worldview associated with it. So it is important to study the religiosity of individuals and societies. Individual political decision makers have their own religious convictions, however, it seems that they are forced to take religion into account in a degree appropriate to the internal cultural and/or ideological conditions of a given country. This applies both to democratic systems in which power comes from choice and represents the collective sovereign, that is, the nation, the people of the country and to non-democratic systems, which also have to take religion into account, especially if it has considerable social power. Much depends on the type of state–religion relationship in force in a given country. On the other hand, it is also clear that atheistic political systems force politicians to diminish the importance of religion in society or even fight against it.

The basis of research into religiosity is primarily self-declaration, usually in answer to a question concerning the role played by religion in the respondent's life. Involvement in religious practices is also measured in such research, although the processes of the privatization of religion are somewhat changing the importance of this factor. Usually, however, participation in services of worship as an external expression of religious adherence is taken as the basis by which the religiosity of a given community, group or society can be determined. In what follows, I primarily refer to research based on self-declaration.

In a Gallup survey of 114 countries and territories in 2009, an average of 84 per cent of adult respondents stated that religion plays an important role in their lives. Most religious (98 per cent and more positive answers) are the Islamic states of Bangladesh, Niger, Yemen, Indonesia, Malawi, Sri Lanka, the Somaliland region, Djibouti, Mauritania and Burundi (Crabtree 2010). This research confirms the rule that the lower the national income per capita, the greater the religiosity (in areas with incomes up to US$2,000 on average, as many as 95 per cent of the population claim that religion is an important part of their lives, from US$2,001 to US$5,000 – 92 per cent, from US$5,001 to US$12,500 – 82 per cent, from US$12,501 to US$25,000 – 70 per cent claim that religion is an important part of their lives, and when incomes are over US$25,001 those who do not consider religion to be an important part of their lives are in the majority, by 52 to 47 per cent). The United States is a rare exception to this rule as 65 per cent of Americans say that religion is important to them. Also, most Italians, Greeks, Singaporeans and inhabitants of the wealthy Gulf states consider religion to be important. Among the least religious are the populations of Estonia (16 per cent think religion is important), Sweden (17 per cent), Denmark (19 per cent), Japan and Hong Kong (24 per cent each), the United Kingdom (27 per cent), Vietnam and France (30 per cent each), Russia and Belarus (34 per cent each think religion is important). Regrettably, many countries which are traditionally among the least religious, such as the Czech Republic and China, were not included in this research. Nevertheless, it clearly shows the very high religiosity of countries in Africa, the Middle East and Latin America, as well as Central and Eastern Europe (Crabtree 2010).

According to a 2015 WIN/Gallup survey conducted in sixty-five countries (this time also in the Czech Republic and China), 63 per cent on average of all the surveyed populations considered themselves religious, and 22 per cent non-religious, 11 per cent declared themselves to be atheists (and 4 per cent did not respond). According to this research, only 56 per cent of Americans claim to be religious. The highest average level of religiosity (86 per cent) can be seen in African countries and the lowest in Western Europe and Oceania (Gallup International 2015).

Research by PRC indicates a growing level of education (if the number of years spent in education is taken as the indicator) among the adherents of all the world's religions, especially among women, though large disparities still persist. On average, adults over the age of twenty-five have spent 7.7 years at school. The most educated are Jews (13.4 years), then Christians (9.3 years) and religiously unaffiliated (8.8 years). Buddhists are also above the world average with 7.9 years spent at school. Muslims and Hindus (with 5.6 years at school) are decidedly below the world average. The level of scholarization depends to a large extent on the country in which the followers of a particular religion live, but we can assume that religious traditions have a certain significance. Among the adherents of Judaism, Christianity, the unaffiliated and Buddhism, who, as shown above, have more than average levels of scholarization, there are slight disproportions between the time spent in education by men and women (for the latter, it is usually shorter, although among Jews there is no difference at all and in the case of Christians, it is less than half a year). The education of Muslim women is an average of 1.5 years shorter than for men, and for Hindu women it is 2.7 years shorter. Nevertheless, it is the followers of these two religions who have made the greatest progress in the last three generations with regard to the length of time they spend in education. In addition, there has been a simultaneous decrease in the educational gap between men and women (measured in terms of time in education) (Pew Research 2016).

Significant changes relating to religion and religiosity are occurring in particular faiths and geographic regions. We will focus here on the situation in Latin America and sub-Saharan Africa, using PRC reports as our basis. Catholicism in many Latin American countries is losing ground to Protestant denominations, especially evangelical. For example, in Nicaragua, Uruguay and Brazil, as well as among Hispanic Catholics in the United States, the difference between the proportions of those born into Catholicism and adult members of the Catholic Church is over 20 per cent (25, 22, 20 and 22 per cent respectively), which represents a serious challenge to the Catholic Church in a region which has traditionally been considered Catholic. Interestingly, far more Protestants experience deep religious feelings, pray more often, take part in church services and consider religion to be very important in their lives. The differences between Catholics and Protestants in this regard are very large – for example, in Venezuela 10 per cent of Catholics and 49 per cent of Protestants are deeply religious, in Brazil the contrast is 23 per cent versus 60 per cent, in Colombia the contrast is 37 versus 62 per cent, in El Salvador the contrast is 48 versus 71 per cent and in Mexico the contrast is 16 versus 37 per cent (Pew Research Center 2014). These studies suggest that leaving the Catholic Church in Latin America is often driven by the need for a deeper experience of religion, which, as can be seen, is not found in Catholicism to the same degree as in Protestantism.

In sub-Saharan Africa, between 1900 and 2010, the proportion of those following traditional religions fell from 76 per cent of the total population to 13 per cent. The proportion of Muslims increased from 14 to 29 per cent, but the largest increase was recorded among Christians – from 9 to 57 per cent, which based on absolute numbers represents an increase of nearly seventy times – from 7 to 470 million (Pew Research Center 2010). Never, in any region of the world, has the growth of Christianity been so great, and it was obviously linked to an enormous rise in the birth rate, but also to effective missionary activity. PRC research confirms the deep attachment of the region's inhabitants to religion, which has traditionally been the 'cement' holding together every part of their lives. Both Christians and Muslims acknowledge the authority of their sacred books and participate on a mass scale in religious practices. The vast majority pray at least once a day and rarely abandon the religion into which they were born. In the context of the conflict between the two religions in the region, and although Christianity is developing more dynamically, on average 43 per cent of sub-Saharan Christians see Islam as violent and only 20 per cent of Muslims view Christianity in the same way. On average, 28 per cent of the people in the region perceive the Christian–Muslim conflict as a major problem for their country, a view which is more often held by Muslims than Christians (32 and 26 per cent respectively). On average, 25 per cent of the population still practise certain beliefs derived from traditional religions (especially belief in the protective role of ancestors and spirits, 'the evil eye' and reincarnation), although the distribution is very varied – from 62 per cent in Tanzania to 3 per cent in Rwanda (Pew Research Center 2010).

Increasingly accurate data concerning religion and religiosity in the world allows us to exclude the thesis of the decline of the importance of religion in human life, an idea which has been widely disseminated in the social sciences. The development of science and technology have not eliminated the religious worldview. A high level of prosperity has affected the degree of religiosity essentially only in Europe, which seems an 'island of secularism' in an ocean of faith. Undoubtedly, in different regions of the world, big changes in religiosity are indeed taking place, but outside of the Old Continent, they are not indicative of the progress of secularization – on the contrary, religions, especially Islam and Christianity, are developing dynamically in terms of the number of followers and their religious zeal. The future of religion will be largely determined by the religious situation in Asia, the world's most populous continent and especially in the still officially 'non-religious' China. However, other regions of the non-European world, such as Africa in particular, are becoming increasingly important.

'The ambivalence of the sacred' in international relations

The concept of 'the ambivalence of the sacred' has become popular in the study of international relations as it points to both the functional and dysfunctional potential of religion in social relations. Individuals (e.g. Desmond Tutu, John Paul II and the Dalai Lama) and religiously motivated groups assist in the alleviation of tensions and conflicts and the development of peaceful relations between the parties in dispute. They contribute to the work of international organizations, intervene on behalf of the most needy, advocate for the protection of human rights and are involved in the process of creating more humanitarian concepts of development. On the other hand, religious motivations and the distinctions which these bring about among people, fuel divisions within and between countries. Religion strengthens the identity and loyalty of rival groups, thus becoming a source of conflict and acts of violence. It is worth noting that religion can directly set political goals (Kulska 2013: 90), especially when religious institutions are closely linked to governments. Furthermore, members of various religions, most often Christians, who constitute one third of the world's population, are discriminated against and persecuted in different countries. The Organization of Islamic Cooperation condemns 'Islamophobia' and the Jews draw attention to anti-Semitism, which is still present in the world in various forms. But as it turns out, not only atheism in totalitarian systems and secularism in liberal systems, but also religion in its ideologized form has been and is being used against people. Proof of this is the phenomenon of terrorism in the name of religion, which has become one of the greatest security challenges in the twenty-first century. This raises questions as to the real intentions of the parties to 'the conflict between the religions', and the real role of religion in the world.

Even before 11 September 2001, R. Scott Appleby, author of the concept of 'the ambivalence of the sacred' cites (critically) in the Introduction to his book the opinion that: 'Religion is powerful medicine, and it should be administered in small doses, if at all' (Appleby 2000). Given the strength of religious attachment, especially in the populations of the global South, religion seems an easy and effective tool to use for achieving political goals. 'Warriors of God' filled with a sense of mission to spread their faith at all costs, have appeared over the centuries in almost every religious tradition. However, it should be remembered that this tool can easily get out of control. On the other hand, religion can also be an inspiration for reconciliation and peace because, as the representatives of the religions themselves, along with academics who are positively inclined towards them, point out, the primary function of religion is to search for a structure by which we can relate to the divine and/or ourselves and consequently to other people. So those who commit crimes in the name of religion are not its true followers.[15]

An extremely important aspect of this peculiar ambivalence is the place of the religious factor within a given political system, which varies according to region but above all tradition and culture. Theoretical treatments of this subject frequently fail to take account of the findings of empirical research. Jonathan Fox has analysed data relating to the religious policy of the state in 177 countries with specific regard to the official establishment of religion (which means that the state adopts a particular religious viewpoint as officially binding), legal and financial support given by the government to believers, government regulation of the majority religion and discrimination against religious minorities.

Fox's research shows that every country has a specific religious policy, unique to itself – there are no two countries which are identical in this respect. In sum, 126 out of 177 countries surveyed (71.2 per cent) have general frameworks supportive of religion, sixteen countries (9 per cent) can be classified as hostile to religion and thirty-five (19.8 per cent) as in the main religiously neutral. This means that the overwhelming majority of the world's countries support religion. It is also interesting that among the countries associated with Christianity, Western and former Eastern bloc countries more often intervene in favour of their religion than do the Christian countries of the Third World. Countries with Muslim majorities more often have an established religion and are more involved in official support for and regulation of their religion, and also more frequently discriminate against minority religions than do non-Muslim majority states. State support for religion is greatest in Muslim majority states in the Middle East and non-Soviet Asia, and least in communist states, Third World Christian states and states with no majority religion (but only South Africa does not use any of the 51 forms of support listed in the research, and only Albania uses just one of them) (Fox 2013).

Religion is regulated least in countries without a majority religion and in Third World Christian states. The highest level of regulation (i.e. the greatest state intervention) occurs in communist[16] and Muslim majority states, but not those in sub-Saharan Africa. Only thirty-one states (17.5 per cent) do not practise any form of religious discrimination. It is also interesting that Third World Christian majority states are closer to the separation of religion and state than those outside the Third World, including the liberal Western states. Catholic countries are more likely to support religion than non-Catholic, except for those with an Orthodox Christian majority. Fox notes that this data does not confirm the stereotype of Arab states as the Muslim countries which are least tolerant towards other religions. Non-Soviet Asian Muslim majority states provide the strongest support to Islam and are the most repressive towards religious minorities. In every religious group, sub-Saharan African countries have the highest level of separation of religion from the state. Former Soviet Muslim majority states are exceptions to the general tendency in Muslim countries with regard to state support for a single religion and discrimination against minorities. Western democracies do not have the highest degree of separation of religion from the state, and neither do they have the lowest level of discrimination, which, as Fox notes, is also contrary to stereotype (Fox 2013).

An important determinant of the functional or dysfunctional character of 'holiness' in social relations, including on the international plane, is political theology which to discuss in relation to the different religions would take us beyond the boundaries of this chapter. It is sufficient to emphasize here that it is not only religious ideas but also religious institutions that determine whether the activity of a given actor has a positive or negative effect on world politics (Kulska 2013: 89).

The study of international relations has a long tradition of seeing religion as a source of conflict. This is linked to the belief that the Treaty of Wesphalia, concluded between the European states in 1648, 'secularized' international relations and so reduced the importance of the religious factor to the internal sphere of the state. According to Michael Barnett, this is a 'legend' which unjustly persists in academic circles, since, in fact, the liberal international order from the seventeenth century onwards was not devoid of religious underpinnings, and many activities of states in the field of foreign policy were based on religion, for example, support for religious freedom (Transatlantic Academy 2015: 23). The opening words of the Treaty of Vienna of 1815, which constituted the basis of a new order in Europe, are an invocation of the most Holy Trinity, thus undoubtedly expressing the common Christian values at least nominally present in the policies of the European states of the time (Banchoff 2015). Barnett argues that 'religion is not a leading cause of violence', rather it is often simply demonized. After all, it was not responsible for the great crimes committed during the First World War and the Second World War, or the genocides in Cambodia and Rwanda. Contrary to appearances, it did not play a very significant role in the Balkan wars of the 1990s, nor in the decades-long Arab–Israeli conflict (Transatlantic Academy 2015: 25). However, it cannot be denied that after 11 September 2001, terrorism in the name of religion has become a predominant model of violence and religious extremism has become a central issue in international affairs (Kulska 2013). Studies also point to the growing importance of religion in civil wars. In forty-four out of 133 such conflicts, in the period 1940–2010, religion was an underlying factor, although it was not always the central cause. The share of religious wars in all internal conflicts since the 1970s has increased visibly – in 2010, they accounted for half of conflicts of this type. Islam was involved in about 80 per cent of such wars throughout the period under review. According to these findings, religious wars last longer and are more difficult to bring to an end, are more likely to resumed and cause more casualties among civilians (Toft 2012; Shah 2011; Toft *et al.* 2012; Kulska 2013: 90). Terrorist acts motivated by religion were on the increase from the 1970s onwards, although it was 11 September 2001, which inaugurated a new and marked upward trend. Particularly since 2005, there has been a dramatic increase in the number of religiously motivated suicide attacks. Between 1998 and 2004, 991 terrorist incidents out of 4824 (or 20.5 per cent) had a religious perpetrator. Islamic-based ideology was the cause of 972 out of the 991 cases (98.1 per cent).

Terrorist acts with religion as a factor have been noted in thirty-seven countries (Toft 2012: 139).

Increasing attention is being paid in the literature on international relations to the positive influence of the religious factor in this sphere of social relations, although the awareness of its negative influence still seems to predominate. But the religious factor is not only a cause of conflicts, it also facilitates their resolution, and contributes to the building of peace and reconciliation after a confrontation has ended. The involvement of religiously motivated ('faith-based') actors, both individuals and groups, in peace initiatives has a very long tradition in many world religions. In Christian Europe, in the second half of the eleventh century, armed conflict was forbidden on certain days of the week and during certain liturgical periods (this was known as 'Treuga Dei' or 'Truce of God'). In addition, popes and people of the Church (e.g. St. Francis of Assisi) exercised their authority in order to bring about the settlement of conflicts, including interreligious wars. In the Middle Ages, Moses Maimonides (Rambam), one of the greatest Jewish scholars, established the ritual of *teshuva*, or repentance which brings about reconciliation with people and God. The traditional Islamic rituals of *sulh* and *musalaha*, signifying peace and reconciliation between hostile groups, and an end to the vicious cycle of vengeance, are of particular importance in the context of the clan feuds which characterize Muslim culture (Toft *et al.* 2011: 178; Kulska 2013: 95).

In modern times, we can also find many examples of religiously motivated attempts to bring about reconciliation both by individuals (Mahatma Gandhi, Martin Luther King, all the popes of the twentieth and twenty-first centuries, the Dalai Lama, Desmond Tutu, and many others) as well as groups (e.g. the Quakers during the conflict in Nigeria between 1967 and 1970, the World Council of Churches and the All Africa Conference of Churches during the Sudan War of 1972, and the Community of Sant'Egidio in Mozambique in 1992). Religiously motivated mediators have a number of advantages from the perspective of the parties to a conflict: they are not guided by specific interests and are not dependent on the success of the negotiations; and they are perceived as impartial and objective, have time and often also sufficient human and financial resources (Kulska 2013: 98). They are usually well oriented in the local aspects of the conflict and are able to find the appropriate route to a resolution. All this makes them effective, especially in internal conflicts, and they also contribute greatly to reconciliation and 'transitory justice' (i.e. in transitional periods, organizing, for example, truth and reconciliation commissions) when the conflict is over (Kulska 2013: 96). In twenty-six selected instances of peace negotiations since the end of the Cold War, an important role has been played by religious mediators in eleven cases, a weaker role in another eleven, and only in four cases were religious mediators absent. In nineteen cases of 'transitory justice' analysed, religious actors played a key, most often direct, role in eight of them (Toft *et al.* 2011: 188–189; Kulska 2013: 98).

In the international arena, the efforts of religiously motivated individuals and groups has found expression in a wide range of undertakings, including the struggle to ban slavery and the slave trade (the United Kingdom and the United States), and the founding of humanitarian organizations such as the International Committee of the Red Cross (established by the Swiss Calvinist Henry Dunant). President Woodrow Wilson's Presbyterian faith inspired him to propose the founding of the League of Nations, and similarly many Protestant churches in the United States supported the establishment of the United Nations. The institutionalization of international cooperation was supported by the Holy See. Many religiously motivated people from different cultural backgrounds – Christian, Confucian, Hindu, Jewish and Muslim – worked with Eleanor Roosevelt on the Universal Declaration of Human Rights. Satyagraha Mahatma Gandhi who led a peaceful struggle against occupying forces and Martin Luther King who inspired the American civil rights movement were also religiously motivated. The European federalism which led to the emergence of the European Union was a movement to a significant degree motivated by the Catholic vision of peace and supported by the authority of Pius XII. There are a number of other examples in modern times of the active involvement of Catholic popes in the quest for peace, including Benedict XV's unsuccessful appeal for the First World War to be halted, the mediation of John XXIII during the Cuban crisis of 1962 and John Paul II in both the British–Argentinian war over the Falklands in 1982 and the 1984 conflict between Argentina and Chile over the Beagle Channel. The Islamic world responded positively to the attitude of the Polish Pope during the Gulf Wars of 1991 and 2003, which meant that those conflicts could not be regarded as clashes of civilization. Many examples of the positive influence of the religious factor on conflict resolution and peacemaking can also be found in the Islamic and Buddhist traditions (Toft *et al.* 2011).

Conclusion

As sociologist Anthony Giddens observes, throughout human history 'religion has continued to be a central part of human experience, influencing how we perceive and react to the environments in which we live' (Giddens 2009: 676). This fact cannot fail to have an impact on international reality. Undoubtedly, between a religious worldview and the rationalist perspective, which dominates many spheres of our lives, there is as Giddens puts it 'an uneasy state of tension' (Giddens 2009: 676), but it is not excluded that a completely new quantum approach to the social sciences will significantly reduce this. For example, the quantum concept of consciousness, which does not overturn the rational understanding of reality but extends it to other dimensions, turns out to be close to ideas which have for centuries been taught by the religions. Perhaps the reason the phenomenon of religion persists in the world is simply because it best answers the questions of the meaning and purpose of our existence (cf. Giddens 2009: 676), and this is something which social science must not overlook. As experience shows, many people have no difficulty in reconciling religiosity with pragmatism and rationalism.

The influence of religion on international relations is different in different regions of the world, because the religiosity of people varies, as does the role which religion plays in the social and political life of societies. However, there is no indication of a decline in the importance of the religious factor in the world, especially in the global South, that is, in Asia, Africa and Latin America, although religions and religiosity in those places are also undergoing numerous changes. There is no reason to expect a sudden increase in religiosity in the affluent states of the West (except for United States which remains religious), although there are some indications of an increase in the presence of religion in the public sphere, for example in secular France.[17] The diverse views and values represented in the human family are encountering and engaging each other at the level of the international system due to globalization's acceleration in recent years. It is worth noting, with Thomas Banchoff, that the most important actors in the international system – national states, market economies and international institutions – operate in a 'strikingly secular' (cf. Banchoff 2015) environment, and there is little to suggest that any alterations will be made in the near future to the systemic operating principles adopted after the Second World War, which are devoid of reference to religion. But it is also certain that the religions, especially Islam and Christianity, have not yet said their last word, as indicated by, among others, the demographic shifts taking place in the world (Banchoff 2015). Religion will, therefore, remain an important factor of change on the political map of the world and will also shape the 'new geographies'.

Acknowledgement

The article was written during a research stay at the Mershon Center for International Security Studies (the OSU) and I am grateful to all for any help.

Notes

1 The full quote from Fromm in context shows that he is explicitly talking about religion 'in a broad sense' which includes non-religious systems of thought, sometimes called 'para-religion', which is not clear in some uses of this quote (thanks to Mark Zniderch for sharping this point for me – A.M.S.).
2 This term was coined by Thomas Henry Huxley, an English biologist and advocate of Darwinian theory, who in 1869, so defined his own religious beliefs (Johnson and Grim 2013: 27).
3 Atheists reject the idea of the existence of any 'divinity', and usually opposed to theism and all forms of institutionalized religion. They often include advocates of ideologies which fight against religion – communists, materialists, Maoists and Marxists. The first country in the world which was officially atheist was, until 1991, Albania. Today, atheism is promoted in bestselling books especially by Sam Harris, Daniel Dennet, Richard Dawkins and Christopher Hitchens (Johnson and Grim 2013: 41).
4 Observers of religious life in China point to the country's enormous potential number of religious adherents, who may be concealing their beliefs due to the continuing repression against various religions, thus giving rise to errors in the statistics (Johnson 2017: 83–95).

5 Atheism and lack of religious affiliation in the modern world has been the subject of in-depth research by Phil Zuckerman in particular (Zuckerman 2007: 47–65, 2009; Solarz 2012).

6 This group includes not only those who recognize Trinitarian baptism, which is the basis of ecumenical cooperation, cf. the 'Lima Document' titled 'Baptism, Eucharist and Ministry', and believe in the Resurrection, but also all who in some way claim adherence to the person of Jesus Christ, for example, Jehovah's Witnesses and Mormons.

7 The authors of the report include those who, for example, do not believe in the Trinity, or who recognize other sources of Revelation beyond the Bible, and so, among others, the Jehovah's Witnesses and Mormons mentioned in Note 6.

8 For the same year, 2010, T.M. Johnson and B.J. Grim have the countries in a slightly different order and with fewer Muslims in some of the countries: Indonesia (190.5 million), India (173.3 million), Pakistan (nearly 167 million), Bangladesh (134 million), Iran (73 million), Egypt (72.4 million), Nigeria (72 million), Turkey (71.5 million), Algeria (nearly 35 million) and Morocco (31.8 million) (Johnson and Grim 2013: 21).

9 The word can then be translated as 'a being set upon enlightenment'. This is a human being 'committed to the attainment of enlightenment for the sake of others' and becoming a bodhisattva is the goal of Mahayana Buddhism (cf. Wildmind n.d.).

10 In 2010, 6 million Jews lived in Asia, 5.6 million in North America, nearly 2 million in Europe, nearly 1 million in Latin America, about 130,000 in Africa and 117,000 in Oceania (mainly Australia and New Zealand) (Johnson and Grim 2013: 51).

11 It should be noted that the data cited above relates to Confucianism as a religion, mainly in South Korea (where 82 per cent of all adherents of this religion lived in 2010), as well as in Myanmar, Thailand, Japan and some other countries. Confucianism as a philosophical system is by many identified with the 'Chinese way of life', and its influence in China and throughout East Asia is enormous (Johnson and Grim 2013: 64).

12 Slightly different data (4 million) is given by Adherents.com.

13 WRD gives the much lower number of 197,000 (Johnson and Grim 2013: 72).

14 More information on this subject can be found here: (The Facts and Stats).

15 Such a vision is presented by, for example, the Catholic Church in papal messages on such occasions as the World Day of Peace. The conciliatory function of Islam is defended by the Organization of Islamic Cooperation and Islamic religious authorities, including from Egypt, the United States, Canada, Pakistan and Indonesia, for example, in their letter to Islamic State leader, Abu Bakr al-Baghdadi (Open letter 2014). However, the function or functions of religion are the subject of endless discussions among sociologists and scholars of religion, although it is generally agreed there are a multiplicity of these (soteriological, worldview, normative, political–ideological, integrative, etc.). They are often discussed in connection with the equally debatable issues of the genesis and definition of the phenomenon of religion. Niklas Luhmann, in contrast to many other scholars, describes the functions of religion in the singular, that is he reduces all its functions to one, which he says is the transformative function of 'representing the non-representable'. Luhmann considers religion as a kind of language which, through faith, transforms the 'inexpressible' into interpretations which carry meaning (cf. I Borowik, *Introduction* in Luhmann 2007).

16 Fox categorizes the following countries as communist: China, Cuba and North Korea (Fox 2013: 200).

17 Especially if the Jesuit-educated François Fillon wins the French presidential election in 2017 (in June 2017, we know that he failed – A.M.S.). It seems that since the 2013 legalization of homosexual 'marriage' with the possibility of adopting children, and especially in view of the growing wave of Islamic terrorism, the slogan of a return to France's Christian roots is gaining increasing traction in French society. One of the expressions of this is the emergence of the *Sens commun* (Common Sense) movement (Meichtry and Rocco 2017).

References

Adherents (2005) Retrieved 1 January 2017 from www.adherents.com/Religions_By_ Adherents.html.

Madeleine Albright, M. (2006) *The Mighty and the Almighty: Reflections on America, God, and World Affairs*. New York: Harper Perennial.

Appleby, R.S. (2000) *The Ambivalence of Sacred: Religion, Violence and Reconciliation*. New York: Rowman & Littlefield.

Banchoff, T. (2015) *Religion and World Order*. Retrieved 6 June 2017 from https:// berkleycenter.georgetown.edu/forum/religion-and-world-order.

Barrett, D.B., Kurian, G.T. and Johnson, T.M. (2001) *World Christian Encyclopedia: A Comparative Survey of Churches and Religions in the Modern World*, vol. 1. New York: Oxford University Press.

Bellah, R.N. (1967) Civil Religion in America. *Journal of the American Academy of Arts and Sciences* 96(1): 1–21.

Berger, P.L. (ed.) (1999) *The Desecularization of the World: A Global Overview. Resurgent Religion and World Politics*. Washington, DC: Wm. B. Eerdmans Publishing Co.

Berger, P.L. (1967) *The Sacred Canopy: Elements of Sociological Theory of Religion*. Garden City, NY: Anchor Books.

Bronk, A. (2009) *Podstawy nauk o religii*. Lublin: Towarzystwo Naukowe KUL.

Burgoński, P. and Gierycz, M. (2014) *Religia i polityka: Zarys problematyki*. Warsaw: Elipsa.

Baptism, Eucharist and Ministry (n.d.) (Faith and Order Paper no. 111, the 'Lima text'), World Council of Churchs. Retrieved 12 April 2017 from www.oikoumene.org/en/ resources/documents/commissions/faith-and-order/i-unity-the-church-and-its-mission/ baptism-eucharist-and-ministry-faith-and-order-paper-no-111-the-lima-text.

The Christian Science Monitor (n.d.) Retrieved 2 March 2017 from www.adherents.com/ misc/rel_by_adh_CSM.html.

Crabtree, S. (2010) Religiosity Highest in World's Poorest Countries. Retrieved 2 March 2017 from www.gallup.com/poll/142727/religiosity-highest-world-poorest-nations.aspx.

Curanović, A. (2010) *Czynnik religijny w polityce zagranicznej Federacji Rosyjskiej*. Warsaw: WUW.

The Facts and Stats on '33,000 Denominations'. Retrieved 12 April 2017 from www. philvaz.com/apologetics/a106.htm.

Fox, J. (2013) *An Introduction to Religion and Politics: Theory and Practice*. London: Routledge.

Fox, J. and Sandler, S. (2004) *Bringing Religion into International Relations*. New York: Palgrave Macmillan.

Fox, J. and Sandler, S. (2006) *Religion in World Conflict*. London: Routledge.

Fromm, E. (1950) *Psychoanalysis and Religion*. New Haven, CT: Yale University Press.

Giddens, A. (2009) *Sociology*, 6th edn. Cambridge: Polity Press.

Hackett, C., Stanowski, M., Potančoková, M., Grim, B.J. and Skirbekk, V. (2015) *The Future Size of Religiously Affiliated and Unaffiliated Populations*. Retrieved 6 March 2017 from www.demographic-research.org/volumes/vol. 32/27/32-27.pdf.

Hassner, R.E. (2011) Religion and International Affairs: The State of the Art, in P. Jameds (ed.), *Religion, Identity and Global Governance: Ideas, Evidence and Practice*. Toronto: University of Toronto Press.

Haynes, J. (2006) Religion and International Relations in the 21th Century: Conflict or Cooperation?. *Third World Quarterly* 27(3): 535–541.

Haynes, J. (2007) *An Introduction to International Relations and Religion*, Harlow: Pearson Longman.

Hoover, D.R. and Johnston, D.M. (eds) (2012) *Religion and Foreign Affairs: Essential Readings*. Waco, TX: Baylor University Press.

Horodeński, A. (2012) *Śledztwo w sprawie Pana Boga*. Warsaw: WAB.

Huntington, S.P. (1993) The Clash of Civilizations?. *Foreign Affairs* Summer issue 72(3).

Huntington, S.P. (1996) *The Clash of Civilizations and Remaking of the World Order*. New York: Simon & Schuster.

Jenkins, Ph. (2009) *Chrześcijaństwo przyszłości: Nadejście globalnej Christianitas [The Next Christendom: The Coming of Global Christianity]*. Warsaw: Verbinum.

Johnston, D. (ed.) (2003) *Faith-Based Diplomacy: Trumping Realpolitik*. New York: Oxford University Press.

Johnston, D. and Sampson, C. (eds) (1994) *Religion, the Missing Dimension of Statecraft*. New York: Oxford University Press.

Johnson, I. (2017) China's Great Awakening: How the People's Republic Got to Religion. *Foreign Affairs* 2.

Johnson, T.M. and Grim, B.J. (2013) *The World's Religions in Figures: An Introduction to International Religious Demography*. Malden, MA: Wiley-Blackwell.

Jurgensmeyer, M. (1994) *The New Cold War?: Religious Nationalism Confronts the Secular State*. Berkeley, CA: University of California Press.

Jurgensmeyer, M. (2003) *Terror in the Mind of God: The Global Rise of Religious Violence*. Berkeley, CA: University of California Press.

Juergensmeyer, M. (2006) *The Oxford Handbook of Global Religions*. Oxford: Oxford University Press.

Jurgensmeyer, M. (2008) *Global Rebellion: Religious Challenges to the Secular State, from Christian Militias to al-Qaeda*. Berkeley, CA: University of California Press.

Kulska, J. (2013) 'Ambiwalencja świętości' jako przejaw czynnika religijnego w stosunkach międzynarodowych. *Stosunki Międzynarodowe – International Relations* 2: 85–100.

Luhmann, N. (2007) *Funkcja religii*. Kraków: NOMOS.

Maps of the World (n.d.) Top Ten Countries with Largest Muslim Population. Retrieved 8 March 2017 from www.mapsofworld.com/world-top-ten/world-top-ten-countries-with-largest-muslim-populations-map.html.

Meichtry, S. and Rocca, F.X. (2017, 3 January) François Fillon, Embracing his Catholicism, Challenge France's Secular Tradition, *The Wall Street Journal*. Retrieved 6 June 2017 from http://wwrn.org/articles/46526/.

Pew Research Center (2010) Tolerance and Tension. Islam and Christianity in Sub-Saharan Africa. Retrieved 6 May 2017 from www.pewforum.org/2010/04/15/executive-summary-islam-and-christianity-in-sub-saharan-africa/.

Pew Research Center (2011) Global Christianity: A Report on the Size and Distribution of the World's Christian Population. Retrieved 4 April 2017 from www.pewforum.org/2011/12/19/global-christianity-exec/.

Pew Research Center (2012) The Global Religious Landscape. A Report on the Size and Distribution of the World's Major Religious Groups as of 2010, December 2012. Retrieved 26 May 2017 from www.pewforum.org/2012/12/18/global-religious-landscape-exec/.

Pew Research Center (2014) Religion in Latin America: Widespread Change in a Historically Catholic Region. Retrieved 2 March 2017 from www.pewforum.org/2014/11/13/religion-in-latin-america/.

Pew Research Center (2015a) 10 Countries With the Largest Muslim Populations, 2010 and 2050. Retrieved 12 April 2017 from www.pewforum.org/2015/04/02/muslims/pf_15-04-02_projectionstables74/.

Pew Research Center (2015b) The Future of World Religions: Population Growth Projections 2010–2050. Why Muslims are Rising Fastest and the Unaffiliated are Shrinking as a Share of the World's Population. Retrieved 12 May 2017 from www.pewforum.org/2015/04/02/religious-projections-2010-2050/.

Pew Research Center (2015c) U.S. Public Becoming Less Religious, Modest Drop in Overall Rates of Belief and Practice, but Religiously Affiliated Americans are as Observant as Before. Retrieved 23 April from www.pewforum.org/2015/11/03/u-s-public-becoming-less-religious/.

Pew Research Center (2016) Religion and Education Around the World: Large Gaps in Education Levels Persist, But All Faiths are Making Gains – Particularly Among Women. Retrieved 7 May 2017 from www.pewforum.org/2016/12/13/religion-and-education-around-the-world/.

Pew Research Center (2017) The Changing Global Religious Landscape. Retrieved 6 March 2017 from www.pewforum.org/2017/04/05/the-changing-global-religious-landscape/?utm_source=Pew+Research+Center&utm_campaign=35571839cb-EMAIL_CAMPAIGN_2017_04_05&utm_medium=email&utm_term=0_3e953b9b70-35571839cb-399972993.

Philpott, D. (2007) Explaining the Political Ambivalence of Religion. *The American Political Science Review* 101(3): 505–525.

Prothero, S. (2010) *God is Not One: The Eight Rival Religions that Run the World – and Why Their Differences Matter*. New York: HarperOne.

Sandal, N.A. and Fox, J. (2013) *Religion in International Relations Theory: Interactions and Possibilities*. London: Routledge.

Snyder, J. (ed.) (2011) *Religion and International Relations Theory*. New York: Columbia University Press.

Shah, T.S. (2011) God and Terror. Retrieved 5 April 2017 from www.thepublicdiscourse.com/2011/05/3316/.

Shah, T.S., Stepan, A. and Toft, M.D. (eds) (2012) *Rethinking Religion and World Affairs*. New York: Oxford University Press.

Solarz, A.M. (2012) Religia w Trzecim Świecie, in M.W. Solarz (ed.), *Kraje rozwijające się na początku XXI w.* Warsaw: WUW.

Telhami, S. and Barnett, M. (eds) (2002) *Identity and Foreign Policy in the Middle East*. Ithaca, NY: Cornell University Press.

Thomas, S.M. (2005) *The Global Resurgence of Religion and the Transformation of International Relations: The Struggle for the Soul of the Twenty-First Century*. New York: Palgrave.

Toft, M.D. (2012) Religion, Terrorism, and Civil War, in T.S. Shah, A. Stepan and M.D. Toft (eds), *Rethinking Religion and World Affairs*. New York: Oxford University Press.

Toft, M.D., Philpott, D. and Shah, T.S. (2011) *God's Century: Resurgent Religion and Global Politics*. New York: W.W. Norton & Company.

Transatlantic Academy (2015) Faith, Freedom, and Foreign Policy. Challenges for the Transatlantic Community, April. Retrieved 9 March 2017 from www.transatlanticacademy.org/sites/default/files/publications/TA%202015%20report_Apr15_web.pdf.

Warner, C.M. and Walker, S.G. (2011) Thinking about the Role of Religion in Foreign Policy: A Framework for Analysis. *Foreign Policy Analysis* 7: 113–135.

Wendt, A. (2008) *Społeczna teoria stosunków międzynarodowych*. Warsaw: Scholar.

Wendt, A. (2015) *Quantum Mind and Social Science: Unifying Physical and Social Ontology*. Cambridge: Cambridge University Press.

Wildmind (n.d.) What is a Bodhisattva? Retrieved 9 March 2017 from www.wildmind. org/mantras/bodhisattvas.

Zuckerman, Ph. (2007) Atheism: Contemporary Numbers and Patterns, in M. Michael (ed.), *The Cambridge Companion to Atheism*. New York: Cambridge University Press.

Zuckerman, Ph. (2009) Atheism, Secularity, and Well-Being: How the Findings of Social Science Counter Negative Stereotypes and Assumptions. *Sociology Compass* 3(6): 949–971.

7 Endangered Earth

Pollution, resources, global change

Anna Dudek, Jerzy Makowski and
Joanna Miętkiewska-Brynda

One of the characteristics of the twenty-first-century world is the large and rapidly growing pressure on environmental resources due to dynamic population growth. Surging demand for energy (conventional and new energy sources), industrial raw materials (traditional and new) and foodstuffs is a result of the demographic boom, which has also led to the emergence of new needs and the widespread adoption of a variety of spatial behaviours.

The rising pressure on resources is manifested primarily in an intensification and expansion of extraction operations, and the development of new technologies allowing the tapping of resources which until recently have been difficult to access, or unprofitable due to technical difficulties and high acquisition costs (e.g. natural gas and crude oil extracted from shale, and large hydroelectric power plants). A new aspect is that raw materials located on ocean floors and the layers beneath have become the object of exploration and extraction, including, for example, hydrocarbon deposits in the Arctic. This raises the possibility of serious clashes between advocates of environmental protection and interest groups, who lobby for the extraction of raw materials for industry without any regard for the negative environmental consequences this may have.

Unrestricted economic activity is sowing the seeds of potential economic conflicts and political tensions (the right to exploit resources) and even possible armed conflicts. It also carries with it, knowingly or not, a high risk of adverse changes in the natural environment (e.g. extermination of species, fragmentation of habitats, invasive alien species and marine pollution) in many regions of the planet which are still almost pristine (e.g. ocean floors and polar regions).

Unwelcome *newcomers*: invasive alien species

The spread of invasive species constitutes a serious environmental problem. These species are most commonly not native to a given ecosystem and are thus known as invasive alien species (IAS). They are a threat to local species of fauna and flora, occupying their ecological niches, competing with them, feeding on them, transforming their environment and cross-breeding with them (hybridization) (Convention on Biological Diversity 2009). Native species which have enlarged their geographical range or habitat spectrum are also considered invasive

and a threat to biodiversity. It is estimated that since the seventeenth century invasive species have contributed to 40 per cent of animal extinctions (Convention on
Biological Diversity 2009). Invasive species are usually characterized by high
expansiveness, rapid growth and reproduction; they tolerate diverse environmental conditions and easily colonize new areas, often facilitated by human influence. The three most common means of the spread of such species are
international trade and transport, and the deliberate introduction of alien species
for use in agriculture, breeding and pest control (The Nature Conservancy 2016).

The problem of invasive species affects both developing and highly
developed countries. In the former, investment in research is small, and so the
effectiveness of IAS discovery is likely to be lower than in the richer countries.
Therefore, a legitimate conclusion would be that the real number of invasions in
developing countries is proportionately higher. Another explanation assumes
that the number of invasive species is, in fact, relatively lower in less developed
countries, as their economies are more weakly connected to global markets.
Lower degrees of globalization, smaller trade volumes and less developed transport all mean fewer IAS.

The proliferation of invasive species generates enormous environmental, economic and public health costs (Perrings 2005: 14–15). Islands characterized by
high levels of flora and fauna endemism are particularly vulnerable to invasions
of alien species. The expansion of alien species can seriously endanger (or even
lead to the extinction of) rare species with a limited geographical distribution. In
order to prevent this, the governments of island nations often implement biosecurity policies. Restrictive regulations apply, for example, in New Zealand,
where every product (including inorganic goods) brought into the country is
subject to special controls (Biosecurity n.d.). This regime also applies to passenger baggage at ports and airports which undergoes detailed checks. These
security measures have been put into place so as not to repeat the mistakes of the
past when many alien mammal species were introduced into New Zealand by
settlers thereby endangering the country's unique endemic avifauna. In the
Egmont National Park on the North Island, where there is a kiwi (*Apteryx sp.*)
population of 150–200 birds, a programme is underway to eliminate possums
and rats. Possums came to New Zealand with European settlers and they spread
throughout both islands. They compete for food – the native species of flora –
with the local birds whose eggs they also eat, and thus have contributed to their
endangerment. The activity of other species of alien predatory mammals (stoats,
ferrets and cats) has almost led to the extinction of the endemic, flightless kakapo
parrot (*Strigops habroptila*). Currently, there are some 120 of these birds living
on three predator-free offshore islands: Codfish Island, Anchor Island and Little
Barrier Island (Kakapo Recovery n.d.).

A similar security regime is in place on Ecuador's Galapagos Islands which,
almost in their entirety, are a protected area (Galapagos National Park).
However, although the movement of goods and people arriving on the islands
is monitored, the system is not as restrictive as in New Zealand. In the past,
beginning with the discovery of the Galapagos archipelago in the sixteenth

century, various animal species such as goats, dogs and cats were brought to the islands, both intentionally and incidentally, and today they are widespread. Herds of feral goats compete for plant resources with land iguana and Galapagos giant tortoises. Park authorities run programmes to eradicate the goats by shooting them from helicopters. Unsupervised dogs and cats also pose a threat as they hunt for young land and sea iguanas as well as young tortoises and birds. Dogs also transmit canine distemper, which is a risk to sea lions when they come to shore. Therefore, the movement of domestic animals on the inhabited islands should be restricted.

Certain invasive species, such as dogs, can be used to eliminate other undesirable aliens. Since December 2014, Santa Cruz, one of the islands of the archipelago, has had a special unit which trains dogs (conservation dog detection teams) to search for the invasive Giant African Land Snail (GALS). This species is an example of a recent invader – it was first recorded in Santa Cruz in 2010. The GALS is the largest species of land snail, capable of feeding on 500 species of plants; it proliferates rapidly, especially in tropical and subtropical climates (Galapagos Conservancy 2016a). At this stage, the cost of removing the entire population is low, because the infected area is still small – about 20 hectares.

The inadvertent transportation of rats to the Galapagos in the seventeenth and eighteenth centuries led to a decrease in the populations of animals endemic to the islands. Rats hunt for small birds and reptiles and feed on their eggs. Rat eradication programmes focus on the smaller islands (Rábida – 499 hectares and Pinzón – 1815 hectares), because it is believed that only there can the programmes fully achieve their aim. Eradication is carried out by laying down poisoned bait. These are high-risk projects, as the islands also have populations of tortoises, birds, snakes, lizards and invertebrates, which could also be poisoned (Galapagos Conservancy 2016b).

Nature in pieces: the impact of habitat fragmentation

Changes in land use and the reshaping of the natural landscape in order to obtain land for cultivation, urbanization, transportation routes and industry mean that various habitats are becoming increasingly fragmented. This includes forests, wetlands, heathlands and grassland formations. The phenomenon of fragmentation means that a habitat ceases to be continuous, its total surface area is reduced, the resulting fragments are more isolated than previously, and the proportion of habitat edges in the total area is now larger (Pullin 2004).

The consequences of habitat fragmentation can be viewed from the perspective of island biogeography theory, in light of which a fragment of a habitat can be regarded as a kind of island. As the fragment becomes smaller and increasingly isolated, species located on the 'island' gradually decline (become extinct), as do the chances of it being colonized by other species. The result is reduced biodiversity. However, the application of this theory to nature conservation has met with criticism, as the analogy between a marine island and an individual habitat fragment is not perfect (Pullin 2004).

According to researchers, 70 per cent of the world's forest areas are located no more 1 kilometre from the forest edge, and 20 per cent are only 100 metres from it (Haddad *et al.* 2015). This creates a whole array of threats to the plant formations and the populations, which live in those areas.

As already mentioned, fragmentation is accompanied by an increase in the proportion of edges to total habitat area. Edges are very different from the interior of habitats and are often unsuitable for certain species. Microclimatic conditions on the edges of habitats differ from those in the interior. Edge zones are prone to a higher frequency of environmental disasters and are more often penetrated by predatory or competitive species. As the proportion of edges to total area increases, so does the probability of passive emigration from the habitat (Pullin 2004).

Fragmentation of a habitat also means the fragmentation of a specific population of a particular species. Newly emerging, less numerous populations are more vulnerable to extinction, for several reasons. First, such a population may die out due to a random variation in the ratio of births to deaths. Furthermore, in smaller populations, the chances of finding a breeding partner are lower, there is reduced group protection against predators and less efficient group feeding. Social hierarchies and social ties can also collapse more easily. Another threat to small populations is the random loss of genetic variation, and, as a consequence of this phenomenon, the more frequent occurrence of individuals with genetic abnormalities. Small populations are also more vulnerable to environmental catastrophes. Any single unfavourable event (adverse weather, flood, fire, volcanic eruption, etc.) can engulf the entire population and bring about its annihilation (Pullin 2004).

Even after the division of a continuous habitat into several isolated parts, some species can survive in the new environment by migrating between individual fragments. When this occurs, the resulting phenomenon is called a metapopulation. If, however, the fragmentation process intensifies, the moment may come when migration is no longer possible and the metapopulation collapses, which in time leads to the complete disappearance of the species (Pullin 2004).

Fragmentation processes are just one of the many factors which contribute to decreasing population size and biodiversity loss. The effects of fragmentation are still to be fully explored, and the results of contemporary habitat fragmentation processes will only be visible in several decades. This will be of particular importance for the management of protected areas and other areas with remaining habitat fragments (Haddad *et al.* 2015).

Transboundary movements of hazardous waste

The export of hazardous waste from the global North to the less developed South is a problem of planetary proportions. This issue took centre stage in the 1970s when, due to the introduction of rigorous legislation, the costs of the processing and storage of hazardous waste in the developed world rose dramatically. As a result, it was now much cheaper to send the waste to less developed countries.

By the 1980s, it was commonplace for toxic industrial waste to be transported from the global North to countries in Africa or other underdeveloped regions. Unfortunately, these countries usually did not have the technology or infrastructure to store this waste in an environmentally safe way (OECD 2012: 19–20). This situation led to the commencement of work on what became the most important legal document regulating and controlling the movement of hazardous waste – The Basel Convention on the Control of Transboundary Movements of Hazardous Wastes and Their Disposal – which came into force in 1992. It was intended to completely prohibit the export of hazardous waste from the industrialized North to the countries of the global South. However, it very soon turned out that such waste was still being sent to less developed countries by taking advantage of legal loopholes in the document, such as the possibility of such export if the waste was further recycled in the receiving country. It is also important that the United States, uniquely in the highly industrialized world, did not ratify the Basel Convention. The weakness of the regulations contained in the document led to the creation of many regional conventions, such as the Bamako Convention, which prohibits the import of all hazardous waste into African countries. To date it has been ratified by twenty-five of Africa's fifty-four countries (Bamako Convention 1998).

In 1995, an annex to the Basel Convention (known as the Basel Ban Amendment) was proposed in order to prohibit the transportation of waste (including for recycling) from industrialized to less developed countries. For the annex to come into force, it must be ratified by three-quarters of the 185 countries which are party to the Basel Convention, but to date, only eighty-eight have done so (Basel Convention 1995). Nevertheless, the European Union together with Switzerland and Norway have fully incorporated the terms of the Basel Ban in their own legislation.

Used electronics fall into the category of hazardous waste because they contain harmful elements such as brominated flame retardants and heavy metals like lead and mercury. Some countries of the global North take advantage of weaknesses in the international legal system, or other legislative loopholes, to send their own electronic waste to Asian or African countries (e.g. China, India, Pakistan, Indonesia and Ghana) (Figure 7.1).

Wherever in China such waste is stored, virtually all of the local residents are involved in processing it, by stripping out valuable materials such as copper, gold and silver. The best known of these places is the town of Guiyu where electronic waste has been processed since the mid-1990s. Reputed to be the largest electronic waste storage location in the world, its streets, squares and backyards are piled high with computers, cell phones and keyboards (Brigden, Labunska, Santillo and Allsopp 2005). Until recently, most of the waste processed in such places came from Western countries, mainly the United States. Today, however, used electronics from the industrialized countries of South-East Asia (principally Thailand and Malaysia) and from the domestic market in China are starting to predominate (Moskvitch 2012). The use of low-tech methods to recycle electronic waste has catastrophic effects on human health and the environment (Brigden *et al.* 2005).

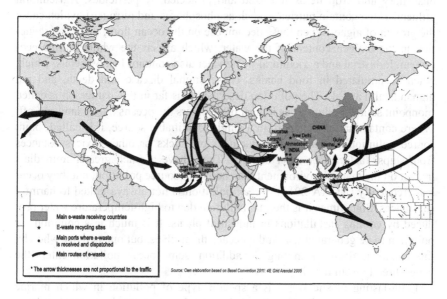

Figure 7.1 The main routes and destinations of electronic waste.

Electronic waste imported into Africa (mainly Ghana and Nigeria, but also Benin, Côte d'Ivoire and Liberia) tends to arrive via a specific procedure and share a characteristic fate. Developed countries send out containers of used computers labelled as 'charitable donations' or 'second-hand goods'. According to a Basel Convention report issued in 2011, the five aforementioned African countries receive 250,000 tonnes of electronic waste every year. About 30 per cent of the electronic devices imported in this way do not work. Half of these can be repaired and are sold on the domestic market, thus constituting an important local source of cheap electronics (Basel Convention 2011). Unfortunately, there are no accurate statistics showing what percentage of these goods end up at waste sites. It should also be remembered that some of the computers, which are still operational when they arrive in Africa, break down quite soon after, and, if they cannot be repaired, are disposed of. The remainder of the imported electronic waste is immediately transported to large rubbish dumps, of which the most infamous is Agbogbloshie in the suburbs of Accra in Ghana. Unprotected sites for electronic waste are becoming a serious threat both to the health of those who live nearby and to natural ecosystems (World Heritage Encyclopedia n.d.).

Murky ocean waters: marine pollution

Pollution enters the oceans from different sources. Most is generated on land as a result of human activity. The principal sources of this type of pollution are municipal, industrial and agricultural effluents. The latter come from animal husbandry and crop fields fertilized and protected by pesticides. Agricultural effluents often contain nutrients (such as phosphorus and nitrogen), which cause the growth of algae. When these decompose on the ocean floor, there is a reduction in the oxygen content of the water, which affects the whole marine ecosystem. Industrial and agricultural wastewater also contains heavy metals, which, when accumulated in food chains, cause animal disease and death, and put human health at risk. The pollution described thus far in this paragraph is called 'nonpoint source pollution' as it most often enters the oceans due to land run-off. Where contamination comes from a single identifiable source, it is called 'point source pollution'. This category includes oil slicks or other toxic substances from ships, factories or other sources. These have a much greater immediate impact on the marine environment than nonpoint source pollution, but they occur less frequently (NOAA 2011). Ocean pollution is not always linked to harmful substances. Sometimes it is thermal pollution due to high-temperature water produced by cooling installations in industrial plants. It is much less common for pollution to be generated within the oceans themselves, but one activity whereby this occurs is deep-sea mining. In addition, some ocean pollution enters the water directly from the air.

Long-lasting plastic waste is a specific type of pollution in which marine animals often become fatally entangled. An additional hazard occurs when large pieces of plastic in the ocean break down over time into smaller fragments due to the action of the sun and the mechanical impact of the waves. Fish, turtles, dolphins and sea birds mistake the plastic pieces for food and swallow them. The largest concentration of plastic waste circulates in the five main ocean gyres (NOAA 2008): the North and South Pacific, the North and South Atlantic, and the Indian Ocean. They have been called 'ocean garbage patches'. The North Pacific Gyre is the largest and most thoroughly researched of these, but the mechanisms by which the waste moves through the gyres are very difficult to investigate. It is known that the gyres are not the ultimate destinations of the plastic waste, but only its fragmentation zones. By various routes, it lands and accumulates on coasts around the world (Nahigyan 2014). Ocean pollution is one of the least studied threats to the environment, which is partly due to the speed and ease with which this type of pollution travels and its varying concentrations at different depths.

As mentioned above, in addition to pollution from external sources, deep-sea mining is a significant threat to marine ecosystems.

It has long been known that the ocean floor hosts raw material deposits. The most thoroughly explored of these are the relatively easily accessible alluvial rock deposits (mainly consisting of sand and gravel) found in ancient river beds in shallow shelf seas, heavy minerals (e.g. cassiterite, magnetite, gold and platinum) and precious stones. The most valuable raw materials, with significant prospects of cost-effective extraction in the near future, include polymetallic concretions (manganese concretions), which in addition to manganese and iron contain more than a dozen other associated elements. Several of these are designated by the European Union as critical raw materials necessary for the development of the technologies of the future (Komisja Europejska 2014). The presence of concretions has been known at least since the HMS Challenger expedition of 1872–1876, although though they attracted no interest for almost 100 years after that. It was not until the 1960s that research vessels under various flags began to explore the oceans in order to discover and evaluate concretion deposits, and undertake the first attempts at extraction.

Practically every ocean has promising mineral deposits and they cover nearly 30 per cent of the surface of abyssal plains (Depowski, Kotliński, Ruhle and Szamałek 1998), with the richest located at depths of 4,000 to 6,000 m.[1] In addition to polymetallic concretions, there are also deposits of Seafloor Massive Sulfides (SEF) resulting from the outflow of highly mineralized waters from hydrothermal vents (including black and white smokers), hydromethane deposits on continental shelves and slopes, and offshore oil and natural gas deposits. Accessing them is extremely attractive, as long as this is not contrary to international law or the principles of environmental protection and sustainable development (see Box 7.1).

A sea area designated as an exclusive economic zone (EEZ) has minimal regulations concerning the exploration, extraction, conservation and management of the natural, living and non-living resources on and below the seabed, and in its waters. EEZ status is regulated by the United Nations Convention on the Law of the Sea (UNCLOS).[2] At times, however, the desire to exercise control over resources and have unfettered access to them by eliminating competition (and also in order to exert political pressure on neighbouring countries) leads to failure to comply with UNCLOS rules.

An example of this type of activity is China's appropriation of the Fiery Cross Reef in the Spratly Archipelago of the South China Sea. UNCLOS does not allow an EEZ to be established around any of the islands of the South China Sea basin. However, this is disputed by the countries of the region – China, Taiwan, Vietnam, Malaysia, the Philippines and Brunei – which occupy dozens of islands in the basin, and China claims the entire Spratly Archipelago. As recently as the first decade of the twenty-first century, the Fiery Cross Reef was the region's most valuable natural site due to its biodiversity and role in the marine ecosystem (Langenheim 2015). In 2014, for the purpose of land reclamation, China began depositing sand and gravel on the reef and covering it with concrete. This led to the complete and irrevocable destruction of the reef areas so treated. John McManus, a marine biologist from the University of Miami who has conducted extensive research in the South China Sea, said that China's reclamation 'constitutes the most rapid rate of permanent loss of coral reef area in human history' (Torode 2015). Countries competing with China in the region are engaged in similar projects, albeit on a smaller scale.[4]

The development of mineral resources located on the ocean floor (or below it) is not a technically straightforward undertaking. For it to be safe, modern and expensive technologies are required, which are still difficult to use and, at the current level of raw material prices, not fully profitable. The commencement of deep-sea mining operations is also not environmentally neutral. Mining will

Box 7.1 Protection versus exploitation

A relatively young category in the area of nature conservation is that of marine and coastal protected areas.[3] The need to protect selected waters became urgent in the last decades of the twentieth century, when existing international agreements proved ineffective barriers to the overexploitation of marine resources, that is, overfishing of the most economically important fish species and undersea mining, an already well-established industry which was now intensifying (Radziejewska and Gruszka 2005).

There are around 5,000 MPAs currently in existence (Protect Planet Ocean 2016). They cover waters around and within archipelagos, as well as coastal plains, river deltas and estuaries, and coastal salt marshes. Typically, MPAs are the living environment for many species of organisms and are extremely important for the reproduction of certain species of fish and crustaceans. The most important objective of MPAs is to protect and preserve representative examples of marine biodiversity for the benefit of future generations (Gormley et al. 2012). This aim is fully implemented almost only in marine reserves ('no-take' MPAs), which represent only 10 per cent of the total area covered by MPAs. In the remaining MPAs, protection is partial as a variety of uses are allowed (multiple-use MPAs). Nevertheless, in regions where MPAs of any type exist (they cover less than 3 million km^2, or less than 1 per cent of the 361 million km^2 of the oceans' surface and about 2 per cent of the 147 million km^2 of exclusive economic zone waters), they are considered to be a reasonably effective way of protecting the ecosystems, which sustain the lives and habitats of marine animals.

The safety of the marine environment is still, however, at risk. The vast bulk of the surface area of seas and oceans (99 per cent) is still not covered by any form of protection. Some MPAs are impressive in size (e.g. the Phoenix Islands in the Republic of Kiribati – 410,000 km^2 and the Great Barrier Reef in Australia – 344,000 km^2), but most have very small areas and over half of MPAs are less than the minimum recommended by scientists (which is between 3 and 13 km^2). This is the situation of the MPAs in the Baltic Sea, which as a whole comply with European Environment Agency (EEA) guidelines – 10 per cent of the Baltic's waters are under protection, but the small size of the individual protected areas, inappropriate location and poor administration undermine the effectiveness of the protection they provide (EEA 2016).

An even greater threat to the marine biodiversity of MPAs is their dangerous proximity to the most important fishing grounds and mineral deposits (Figure 7.2). This situation exists in almost all the world's oceans. With the rapidly increasing pressure on resources, it is difficult to imagine that in these places it will be possible to reconcile the conflicting goals of environmental protection and economic benefit.

Nevertheless, it seems that the creation of marine and coastal protected areas represents a promising opportunity to protect marine biodiversity, and a chance to improve on the current situation in which 'might is right' (see the activities of China in the South China Sea).

No doubt the MPAs are insufficient, and other, more comprehensive tools for the protection of marine ecosystems need to be created. In this field, cooperation between institutions which deal with the oceans is important: both law-making

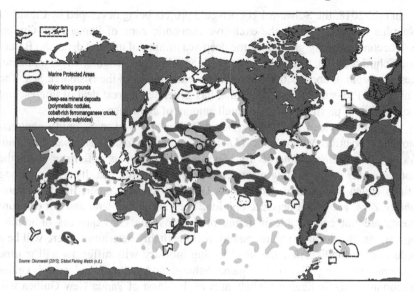

Figure 7.2 Protection versus exploitation.

institutions and those which administer the maritime economy. A beneficial pre-condition for this would be increased public awareness of the effects of human activity on the Earth's marine ecosystems.

affect (or where already underway, is affecting) the ecosystems of the ocean floor, coral reefs and coastal waters in ways and with results which are difficult to predict.

What threats to the natural environment are caused by the mining of submarine mineral resources? There is no simple answer to this question. The impact may range from periodic disturbance of the balance of complex and highly sensitive ecosystems, to the risk of their degradation and annihilation, as in the case of the Fiery Cross Reef. Deep-sea mining activities mean that life in the bottom reaches of the oceans is at risk of accidental mechanical destruction, filter feeding organisms may be harmed by water which has been churned up, and dangers arise when the natural stratification of the water is disturbed by changes in temperature, oxygen content and salinity, and the presence of other chemical compounds (all of which are a potential risk to organisms with a narrow tolerance for environmental factors). The mining of mineral deposits also poses the risk of environmental pollution from various substances,[5] changes in the morphology of the ocean floor caused by underwater landslides and debris flows or collapses of hydrothermal vents. The environmental impact of deep-sea mining may vary depending on the depth and nature of the deposits, the mining technologies used and the resilience of the endangered ecosystems.

In mid-2018, the Solwara 1 copper-gold project being developed by Canada's Nautilus Minerals Inc. in the exclusive economic zone of Papua New Guinea will become a testing ground for the deep-sea mining of mineral deposits (Batker and Schmidt 2015). The copper ore (with an admixture of gold, silver and zinc) lies at a depth of about 1,600 metres in the eastern part of the Bismarck Sea. The Solwara 1 Project, the first of its kind, has generated a great deal of interest and expectation of profit, but also no small amount of controversy.

A report prepared in 2015 by David Batker and Rowan Schmidt, commissioned by Nautilus Minerals Inc. (*Environmental and Social Benchmarking Analysis of the Nautilus Minerals Inc. Solwara 1 Project*), is optimistic about the consequences of the mining of minerals in the Bismarck Sea. Its critics, among others Helen Rosenbaum and Francis Grey (2015) in a report titled *Accountability Zero*, are of the opposite opinion. They argue that land and sea environments are not comparable in terms of the possible consequences of mining operations, that local sea water quality will inevitably deteriorate, there will be a reduction in biodiversity, the local fishing industry will suffer, as will marine tourism, including diving and game fishing, and cultural and social values important to the indigenous inhabitants of the coast of Papua New Guinea will be eroded. Citing the opinion of biologists, Oliver Milman (2012) predicts that deep-sea mining will interfere with natural ecosystems by introducing toxic metals and compounds which may be transmitted though the food chain to organisms caught and consumed by humans.

There are many other countries in this part of the Pacific, such as Fiji, Tonga and New Zealand, which posses resource-rich coastal waters, and they are carefully observing the progress of the Solwara 1 Project. These countries are considering the possibility of granting licenses/concessions to deep-sea mining companies in the hope of securing large revenues (Milman 2012).

In some cases, the intention to access the ocean floor's resources or even assert freedom of navigation, especially in waters whose status is unregulated, is already causing debate and becoming a source of political turmoil. Such an area, incomparably larger than the South China Sea, is the Arctic Ocean with potentially large deposits of oil and natural gas. A dispute is already underway between Russia, Canada, the United States, Norway and Denmark over the scope and boundaries of their exclusive economic zones and the transport routes through and to them. Thus, the Northwest Passage through the Canadian Arctic Archipelago is considered by Canada to be its territorial sea and the Northeast Passage (part of the Northern Sea Route) is claimed by Russia as its territorial sea. Based on the results of shelf research, Russia is also seeking agreement from the *Commission on the Limits of the Continental Shelf* to the expansion of the boundaries of its Arctic zone (The World Factbook 2016). The issue of control over the islands and waters of the Arctic is under negotiation between many parties in light of the provisions of UNCLOS.

Deforestation: the race for raw materials and land

The race for mineral resources and the quest for new energy sources are also taking place on land, which is, after all, where they started. Many different types of activity are involved. We are already used to wind turbines, and no one is surprised by the news of the construction of yet another hydroelectric power plant, or forests being cleared and replaced by plantations used for the production of alcohol, oil or fuel, or for the cosmetics industry. At most, concern is raised by the scale of human intervention in nature, the disregard for irreversible changes caused to the environment, the loss of places of natural beauty and the social costs of these projects. This is an extremely attractive field of inquiry for modern geographic research.

The victims of the unrestrained appetite for resources and energy, and the accompanying destruction and pollution (eutrophication[6]), are whole land ecosystems, such as forests and wetlands, with their complex trophic networks and the environments they occupy in which energy flow and matter cycling take place. Human intervention on land is most often directed at forests, such as the tropical rainforests of the Amazon, Equatorial Africa and the Malay Archipelago.

The clearing of tropical forests, including mangroves, is particularly serious as it inevitably leads to changes (as a rule unfavourable) in the lives of local communities, the extermination of forest-dwelling animals (often already endangered species[7]), and contributes to growth in CO_2 concentrations in the atmosphere, which in turn acts as a further stimulus to global warming.

Slash-and-burn agriculture, large settlement and agricultural colonization programmes, the agro-industry, cattle breeding, firewood and charcoal burning, commercial tree felling and investments in mining, industry and energy production are among the activities responsible for the elimination of forests (Makowski 1999). To these we may add ever more extensive crude oil drilling and urban sprawl. It is difficult to identify the single factor, which contributes most to the devastation of forest areas. Perhaps it is the construction of major hydroelectric plants, as these bring about many other changes in the local environment. Electricity generation can promote the development of other forms of human economic activity, especially large infrastructure projects which attract an influx of migrant workers who find employment in road construction, mining, steel mills and other industrial facilities, or in the agricultural sector, for example, creating and maintaining industrial crop plantations.

The most spectacular example of such activities in a tropical forest zone in the last decades of the twentieth century was the Grande Carajás Programme (in Portuguese: Programa Grande Carajás – PGC) in the eastern part of the Brazilian Amazon. The aim of the programme, which commenced in 1980, was to facilitate the opencast mining of gigantic iron ore deposits (and other metal ores, including bauxite) located in Serra dos Carajás, a low-altitude mountain range. The construction of a large hydroelectric power plant on the Tocantins River was a prerequisite for the mining of the ore, its on-site processing, and the transportation and export of the resulting ore concentrate. Essential components of the project were a railway link connecting the mines and ore-enrichment facilities with the port of São Luís (900 km away), and a road to Belém, capital of the state of Pará, flanked by industrial corridors with steel mills and other industrial facilities. The PGC planning region covered about 900,000 km². When the programme began, 70 per cent of this area was tropical rainforest (Makowski 1999; Enquete Commission 1990).

The total amount of forest which had been cleared under the PGC programme by the beginning of the 1990s is estimated at 250,000 km² (Aragón Vaca 1992). The current situation is uncertain due to the absence of reliable data.

The latest project in the Amazon region with almost unimaginable environmental consequences is the Belo Monte hydroelectric dam complex[8] on the Xingú River, the right tributary of the Amazon. The project was promoted by successive Brazilian governments and large energy companies for nearly forty years before construction finally began in 2011.

The construction of a hydroelectric power plant on the Xingú River was considered as early as 1975. However, it only appeared in the plans of the Brazilian government in 1987 as part of 'Plano 2010',[9] which envisaged the construction of 136 hydro power plants in the country by 2010, most in the tropical rainforest areas of the Amazônia Legal region. Modifications to these plans have meant the extension of the envisaged period of construction to 2030 (and this is the current plan). The largest hydro projects are to be built on the Xingú River – five hydroelectric power stations on a 170 km stretch between the cities of Altamira and Belo Monte. Others are planned on the Araguaia, Tapajós and Tocantins rivers.

The completed part of the Xingú River project – the Belo Monte (initially Kararaô) hydroelectric dam complex is much smaller than in the original plans, mainly due to protests from local people and influential opponents, including church leaders. Under the project, the flow of the Xingú River has been diverted by three dams and numerous dikes, cutting off a huge bend in the river (the 'Volta Grande') and directing approximately 80 per cent of the water, by means of artificial canals, towards the giant turbines of the power plant.[10] The surface area of the artificial reservoir above the dam has been reduced from a planned 1,225 km² to 440 km².

The environmental consequences of the Belo Monte project are not fully known. An area of land measuring 668 km², of which 400 km² is rainforest, is now under water, which is positive to some extent as it represents only a small percentage of what was originally planned for flooding and is incomparably smaller than the area of forests submerged by several other hydroelectric reservoirs built in the Amazon and its vicinity. The periodic (lasting up to seven months a year) but complete dewatering (water blocked upstream by the dam) of a part of the Xingú channel (the Volta Grande), an area of unique value due to its biodiversity and abundance of biocoenoses, represents drastic environmental change.

A serious biological threat is posed by the interruption of the natural migration routes of many fish species (probably also other animal species) due the dams built on the Xingú, the periodic drying out of the Volta Grande and the reduction of biodiversity at all levels: ecosystem, species and genetic. Several endemic species of fish (out of some 600 living in the Xingú basin) are at risk of total extinction (Camargo, Giarrizzo and Isaac 2004; Barbosa Magalhães and del Moral Hernandez 2009). It is estimated that in the first ten years of operation of the Belo Monte hydroelectric power plant, CO_2 emissions from decaying organisms will reach about 12 million tonnes, which will noticeably augment the global greenhouse effect (Richey, Melack, Aufdenkampe, Ballester and Hess 2002; Fearnside 2009).

The social implications of the dam construction on the Xingú River are also significant. More than 20,000 inhabitants of this area will require relocation (Elizondo 2012). These are Indians belonging to a dozen or more ethnic groups, mainly Jurun and Arara. Changes in the environment will also affect their economic behaviour – access to fishing, hunting, agriculture, the use of rivers as transportation routes, and water sources for food and daily hygiene. Standing water in reservoirs can also be an environment conducive to water-borne diseases. Social conflict and a further threat to the environment will be generated by the influx of a potential 100,000 migrant workers into the area, who will settle in the vicinity of the enormous construction project. Their presence and the drain that they cause on local resources will prompt disputes and conflicts with the native population or will exacerbate existing conflicts.

Similar problems seem to exist throughout the tropics, especially in countries with large disparities in development (e.g. Brazil) and underdeveloped countries (most African countries).

Another type of drastic interference in the natural and social environment, which involves increased pressure on resources is the development of plantations of industrial crops, especially oil palm.

Elaeis guineensis, or African oil palm, native to equatorial Africa, is the most productive of the oil plants with a yield many more times that of soya, sunflower or rape. Palm oil extracted from palm fruit pulp (25 per cent of world production of vegetable fats) is cheaper than other vegetable oils and has a very wide range of uses, including as edible fat and an ingredient in cosmetics and lubricants. As much as 85 per cent of the world's palm oil is produced in Indonesia and Malaysia (Prokurat 2013). It is also produced in Thailand and on a much smaller scale in many African countries (DR Congo, Ivory Coast, Ghana, Nigeria and Cameroon) and equatorial America (Brazil, Colombia, Ecuador and Honduras).

The development of plantations and high levels of palm oil production bring many benefits in the form of hundreds of thousands of jobs for local people and development opportunities for many communities, but it also poses serious threats to the natural environment (Górecka 2011). These derive from the unprecedented pace of expansion of monoculture oil palm plantations – driven especially by the largest producers. The spread of plantations means the large-scale extermination of the tropical rainforests. According to the United Nations Environment Programme, if the current rate of palm oil production is maintained, by 2022, we will have seen the destruction of 98 per cent of the rainforests of Sumatra and Borneo (Economist 2010).

The expansion of oil palm plantations is also endangering wetland ecosystems in Indonesia and especially in Sumatra. The decision to drain peatlands and redesignate the land for oil palm plantations has been called by environmental specialists 'a monumental mistake', which will impact Indonesia's long-term prosperity and sustainability (Butler 2009). The draining of peatlands and the fires inevitably used in land clearance will, according to specialist opinion, release huge amounts of CO_2 into the atmosphere.

The destruction of tropical rainforests and peatlands is likely to bring about the extinction of many animal species in the equatorial zone, including anthropoid ape species assigned in the IUCN Red List of Threatened Species to the Critically Endangered (CR) category: orang-utans (The Bornean orang-utan, *Pongo pygmaeus*, and Sumatran orang-utan, *Pongo abelii*) and all species of Malaysian gibbon (the *Hylobatidae* family).

Threats to protected areas

The quest for resources to meet the needs of the world's growing population has resulted in an increasing variety of threats to protected areas. The unsustainable management of resources by local communities is a cause for concern, as are phenomena with external causes, such as predatory forestry, the demand for products made from rare animal species or their body parts, and mineral prospecting by mining companies (tantalum, crude oil, gold and diamonds).

The best example of these concerns is the Democratic Republic of the Congo (DRC), where a significant part of the world's supply of rare earth elements is located, including cassiterite (tin ore), coltan (tantalum ore) and tungsten (also gold and diamonds), and where oil drilling, illegal logging, wild game hunting and the bushmeat trade are booming. Valuable resources are the country's curse. Demand for them is not only fuelling a half-century old armed conflict, but it is also destroying the environment. The exploration and extraction of mineral and other resources is also taking place within the Virunga National Park, the oldest national park in Africa, which is home to a mountain gorilla population (Nellemann, Redmond and Refisch 2010). For many of this region's people, taking advantage of these resources (even illegally) is a matter of survival. Some of the local population believe that the conservation rules implemented by the authorities of the national park are a barrier to their enrichment. Environmental protection is also being undermined by the central government, which has approved plans for oil drilling in the park. Test drilling there has thus far only been conducted by the British company Soco International, but after a wave of international criticism it withdrew from the project (Gouby 2015). The by and large predatory extraction of natural resources, the flouting of environmental norms and disregard for the social consequences of such methods of operation are all facilitated by the unceasing demand for the raw materials used in modern industry – from mobile phone and metalworking tool manufacturers to aircraft and armaments companies (BSR 2010).

The effects are dismal: the forest areas and the habitat of wild animals are shrinking (due to logging, charcoal production and artisanal mining), soil erosion is accelerating, surface waters are being poisoned by the mercury used in gold extraction. The lives of local people are at risk and the tourist trade, which could be an additional source of income for them – an alternative to smuggling – is contracting (Krukowska 2016).

The entire eastern part of the Congo, which is controlled by various rebel groups, is beset by problems relating to the extraction of raw materials. Proceeds from the illegal sale of raw materials – for every 180 kg of legally exported gold, almost 10 tonnes of gold is smuggled across the border (Gatimu 2016) – are used to finance further military operations (BSR 2010). These 'blood minerals' are subsequently found, within electronic devices, in the homes and institutions of the developed world. A solution would be the proper monitoring of the extraction and transportation of raw materials (official data reflects just a fraction of actual production). However, because the DRC is a failed state, this solution is not easy to achieve. The difficulty of tracing the coltan and cassiterite supply chain back from Asian smelting and refining facilities to mines in the DRC makes locating the source of the raw materials a problem.

Rhinoceros horn smuggling (see Box 7.2) and the elephant ivory trade are spectacular examples of the demand for animal body parts. This phenomenon illustrates how certain events in one part of the world can change reality in another, far-distant region.

Elephant hunting reached its highest intensity in the 1970s and 1980s. The largest importer and processor of ivory at that time was Europe (UNEP, CITES, IUCN, TRAFFIC 2013: 64). In order to prevent the extermination of elephants, a global ban on the ivory trade was introduced in 1989. Western societies became very aware of the impact of the fashion for owning objects made of ivory and this led to a decline in demand for these products. The restrictive ban was eased in 1997 with the legalization of the sale of ivory in some African countries (Botswana, Namibia, Zimbabwe and subsequently South Africa). This was subject to the condition that the ivory came exclusively from pre-existing stocks and not from the wild elephant population (UNEP *et al.* 2013: 6). Unfortunately, in recent years, there has been an increase in the number of cases of poaching and ivory smuggling (Milliken 2014: 6). The African elephant population is now estimated to number between 400,000 and 650,000 (UNEP *et al.* 2013: 22). Some 100,000 African elephants, or some 15–25 per cent of the population, were killed between 2010 and 2012 alone (Skinner 2014). Currently, poaching is especially intense in Central and Western Africa, whereas it is relatively less common in the eastern and southern parts of the continent (UNEP *et al.* 2013: 34–35).

The cause of the contemporary increase in demand for ivory is considered to be the growing wealth of countries in the Far East where ivory products are very popular. Smuggling routes begin in the places where elephants are killed and their tusks removed. The ivory is transported by land to ports or airports in West or East Africa; there it is usually placed into ship-borne containers destined for Asia. The sea route is favoured because ivory is heavy and it is most easily hidden in containers among other goods (Milliken 2014: 11). The Asian port of

Box 7.2 Rhino horn smuggling

For centuries, it has been widely believed in South Asia that all parts of the rhinoceros, especially its horn and blood, have miraculous and medicinal properties. Also in East Asia, mainly China, many people are deeply convinced that the rhinoceros horn is an effective medicine for fever, arthritis, typhoid fever, headache, nausea and other ailments. There are no scientific studies confirming the alleged healing properties of rhino horn, but the broad spectrum of efficacy which is claimed for it has ensured that demand has always been very high. In the 1970s and 1980s, rhinos were hunted very intensively, mainly in Africa, with the result that they died in massive numbers and large quantities of rhino horn were exported to South Korea, Taiwan and China (Ellis 2013). In 1989, the efforts of international conservation organizations led to a global ban on the rhinoceros trade. Further assistance in the struggle against intensive poaching came in 1993 when rhinoceros horn was removed from the official list of ingredients used in traditional Chinese medicine. The rhinoceros was placed under strict protection and its populations in conservation areas slowly began to recover. Poaching still took place, but only sporadically (Milliken 2014). It seemed that the international community was bringing this problem under control. But everything changed in 2008 when a rumour began circulating about a former Vietnamese politician who, it was claimed, had been cured of cancer by taking powdered rhino horn. The story became hugely popular and led to a renewed rise in demand for the supposed 'miracle cure'. This is particularly evident in Vietnam, where cancer mortality is currently on the increase, and waiting times for chemotherapy or radiotherapy are long (Guilford 2013). In addition, the high price of rhino horn means that for the Vietnamese the ability to buy it is a symbol of prosperity and status.

According to some specialists, the current demand for rhino horn in Asia, mainly China, is not so much due to its supposed medicinal properties, but primarily because it is a good investment (Gao, Stoner, Lee and Clark 2016).

Therefore, since 2008, we have been faced with a renewed crisis relating to rhino horn smuggling. Annually, hundreds of rhinoceros are illegally killed, both in Africa and Asia. Of the five species of rhinoceros, four are considered endangered, and three of these are critically endangered. Rhinoceros are most numerous in Africa hence most poaching of these animals also takes place there. In South Africa, in 2015 alone, some 1,200 rhinos were killed out of a total population of about 29,000 (International Rhino Foundation n.d.b) (Figure 7.3). Gigantic demand has driven up the price of rhino horn to record levels. The most commonly quoted retail prices are around US$60,000 per kg (Phippen 2016), but they can even reach US$100,000 per kg (Guilford 2013). So it is hardly surprising that attempts to limit poaching are ineffective.

The illegal rhino horn trade is global in scale. The target markets for rhino horn are primarily China and Vietnam, where it arrives mainly by air (see Figure 7.4). However, the effects of the trade reach beyond Africa and Asia. In 2011, very soon after the onset of the current poaching crisis, in many European countries rhino horns were stolen from natural history museums, zoos, private collections and auction houses. Horn smuggling organizations had realized that it was much easier to organize such robberies in Europe than to hunt rhinoceros in Africa. Over the two-year period 2011–2012, Europol recorded sixty-seven thefts and fifteen attempted thefts of rhino horns in Europe (Bailey 2013; Sivarnee 2013).

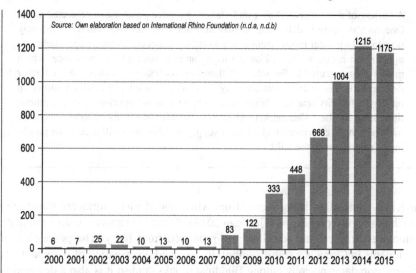

The statistics presented here only relate to South Africa as it is there that the problem of rhinoceros poaching is currently most acute. However, the trends shown in the figures for South Africa also reflect those in other African countries

Figure 7.3 The number of poached rhinos in South Africa, 2000–2015.

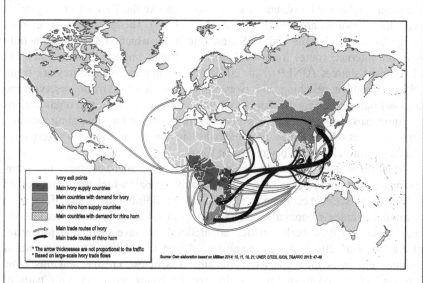

Figure 7.4 The main routes of ivory and rhino horn trafficking.

A variety of measures are being undertaken to reduce poaching in protected areas. One such measure is dehorning, which environmental activists believe is the only way to fully protect rhinos from being killed by poachers. Another approach is to inject rhino horns, while still on the living animal, with a toxic substance, which makes them unsuitable for sale and therefore useless to poachers. As a further deterrent, the horns are simultaneously injected with a pink dye, which shows up on airport security scanners. Work is also ongoing on the development of synthetic rhino horn for the Asian markets. However, this idea has many opponents because not everyone is convinced that a higher supply of rhino horn will reduce the poaching of rhinoceros in the wild.

arrival is most often Hong Kong, from where the elephant tusks are transferred to Chinese ivory processors (Garrigan 2014). These are factories where elaborate sculptures, trinkets and jewellery are produced from ivory. The possession of such items in China is a status symbol. In Thailand, the wearing of ivory amulets is very popular especially among Buddhist monks, and so it is also a destination for African ivory, as are the Philippines, Taiwan and Japan (Milliken 2014: 14) (see Figure 7.4). Today's heightened demand for ivory is not only generated by Asia. The presence of foreign investors (including from China) in the African mining and wood industries, or in large infrastructure projects translates into greater demand in Africa itself (UNEP *et al.* 2013: 11–12). Europe is no longer the main market for ivory, but in some countries like the UK or Germany relatively large amounts are sold (UNEP *et al.* 2013: 64). The second largest market, after China, remains the United States. The ivory which is sold there, mainly over the internet, comes from legal stocks – that is, imported before 1989, but also illegal sources (UNEP *et al.* 2013: 65). A number of measures are being taken to combat poaching and ivory smuggling. Many African reserves and national parks receive extra funds from international agencies and organizations to equip park rangers. There are also severe penalties for smuggling. Mostly, however, it is only ordinary poachers who are caught, and those who organize the illegal trafficking in ivory remain at liberty.

Rhinoceros and elephant poaching attracts most attention from the international community, but these are not the only animals which are being hunted on a massive scale. According to reports on the wildlife trade, some of the most commonly smuggled animals today are pangolins, valued in Asian markets for their meat and their scales, which are believed to have medicinal properties (Heinrich *et al.* 2016). A very significant driver of the illegal trade in various animal species is their use as ingredients in Chinese cuisine and medicine. Among the better-known cases of this are the bones and other body parts of tigers. They are used in China in the preparation of wine with medicinal properties. Tigers are found in different parts of South and East Asia (a population of about 4,000) (WWF 2016), but most (about 5,000) are held in captivity in special breeding farms in China. Another ingredient used in Chinese cuisine is the sea cucumber. Demand for these is so great that they are grown in aquaculture facilities in various countries and then exported to China (Hair, Pickering and Mills 2012). Some species of sea cucumber are even caught in the waters of protected areas, such as the Galapagos Islands Marine Reserve (Toral-Granda 2008). Another Chinese delicacy is edible bird's nest (in Chinese literally 'swallow's nest'), made from the saliva of certain species of swiftlets (*Aerodramus fuciphagus* and *Aerodramus maximus*), which inhabit caves in South East Asia. The nests are used to make a thick soup, which is popular not only in China, but also wherever there is a large Chinese diaspora. In some caves, the gathering of nests takes place after the chicks have hatched, a practice which is considered by the scientific community to be sustainable. This is the practice in the protected Gomantong caves in Sabah, Borneo (Hobbs 2004). Twice a year, licence holders (local individuals) can gather a certain number of nests there. However, methods used in similar caves in the south of Thailand are more dubious. Nest gathering in these caves has been licensed by the authorities to large and influential companies. Access to the caves is very difficult, but all indications are that the nests there are gathered before the chicks have hatched. Demand for edible bird's nest is so great that in some locations such as Kumai in Indonesian Borneo and Sekinchan in Malaysia, local Chinese have invested in the construction of special buildings as artificial habitats for swiftlets to build their nests in (known as swiftlet farming). These buildings resemble windowless houses, inside of which the natural conditions in the caves are recreated. This practice is considered to be environmentally beneficial as it allows the wild bird populations to continue to freely reproduce in caves. In addition, the 'production' of nests in swiftlet farms takes place in more hygienic conditions than in the wild (Thorburn 2014).

The exotic pet trade market is another area of concern. It is dominated by exotic species of birds (e.g. parrots) and reptiles (principally lizards and snakes), and also some mammals (principally monkeys and rodents).

The current rise in interest in exotic animals or their body parts is due to growing purchasing power in countries where luxury goods have hitherto been relatively inaccessible. The trade is also facilitated by widespread access to the Internet through which many such transactions are arranged. Evidence suggests that belief in the medicinal value or miraculous qualities of exotic animal products is declining, and the real driving force is their investment value, the fashion for owning or using certain items, and their importance as indications of wealth.

The scale and effects of trafficking in valuable species, the trade routes used and factors affecting supply and demand, are all very relevant issues for our time, which can be best approached through a combination of the disciplines of geography and biology.

Conclusion

The actual and potential threats to our planet and its inhabitants stem from many factors which are ultimately linked to rising population numbers, growing consumption and increasing pressure on resources in the broadest sense. This pressure forces the search for new solutions and manifests itself in expansion into as yet unexplored regions of the globe. In consequence, the warming of the Earth's climate already seen for several decades, is continuing apace, which in turn is producing phenomena such as the diminishing of the Arctic Ocean's ice cover, glacier recession and permafrost reduction. These phenomena are unlocking access to previously inaccessible resources, and allowing expansion into yet more areas of the planet, leading to further environmental degradation. This will lead gradually to a deterioration in the quality of life, an increase in risks to public health and will even directly endanger the lives of people living in areas increasingly reshaped by human activity (high CO_2 and soot emissions, and rising pollution concentrations in ambient air, rivers, lakes and coastal waters).

The interaction of various processes is exacerbating already existing damage and threats. An example of this is habitat fragmentation, which accelerates the extinction of rare species already decimated by, for example, the expansion of alien species. Both phenomena (fragmentation and alien species invasion) are among the most significant threats to biodiversity in the world today. The mutual and unpredictable interplay of different factors is the result of a positive feedback mechanism. In a globalized world, this manifests itself in the unfavourable correlation of various phenomena and processes, the accumulation of their negative effects, and their ever-widening proliferation. Extremely complex social, cultural and political issues overlap onto, and additionally complicate, environmental problems.

There are no simple and quick remedies for unfavourable changes to the environment. It can be assumed, with Bauman (2007), that environmental problems result from the accumulating negative consequences of how the globalized world functions. We can only hope that humanity will find positive solutions to these problems: the spread of public accountability, good laws and effective systems of justice. However, this will require the cooperation of many actors (social, political and economic).

Notes

1 Since 1991, Poland, Bulgaria, the Czech Republic, Cuba, Russia and Slovakia, within the framework of the Interoceanmetal Joint Organization (IOM) and as members of the International Seabed Authority (ISA), have jointly owned a 75,000 km² section of ocean bed containing polymetallic concretions. The ISA has stipulated that these member states should use the location to conduct geological research and develop seabed extraction technology (www.iom.gov.pl; www.gospodarkamorska.pl).

2 The United Nations Convention on the Law of the Sea (UNCLOS) was signed on 10 December 1982 in Montego Bay, Jamaica.

3 Marine Protected Area (MPA): Marine Reserves, Marine National Monuments, Marine Sanctuaries, Locally Managed Marine Areas and even Marine National Parks.

4 The ruling against China by the *Permanent Court of Arbitration* in The Hague on 12 July 2016 (requested by the Philippine Government in January 2013) did not end the disputes which are ongoing over the South China Sea. It is possible, however, that it has opened a new chapter in the struggle by the great powers for hegemony in these waters (Góralczyk 2016).

5 A prime example of this is the ecological disaster in the Gulf of Mexico and its coastal areas following the oil spill from the Deepwater Horizon platform in April 2010 (Talley 2010).

6 Eutrophication – the process by which a body of water becomes enriched in dissolved nutrients (such as phosphates) that stimulate the growth of aquatic plant life usually resulting in the depletion of dissolved oxygen (Merriam Webster; retrieved 25 May 2017 from www.merriam-webster.com/dictionary/eutrophication).

7 Species assigned to the Endangered (EN) category by the International Union for Conservation of Nature (IUCN).

8 Complexo Hidrelétrico Belo Monte – hydroelectric power plant on the Xingú River in Brazil with a planned capacity of 11,233 MW. When finished, it will be the third largest such facility in the world. The completion of construction and full operation of all turbines is scheduled for 2019.

9 Plano 2010 – Plano Nacional de Energia Elétrica 1987/2010 (Plan 2010 – National Electricity Energy Plan 1987/2010).

10 The now dry stretch of the Xingú channel below the dam was previously a unique place in the Amazon. Here the river flowed slowly, overflowing its banks widely in the rainy season, or, at low water, running in hundreds of channels and streams among countless rocky and alluvial islands which were partly covered with forests, thickets, and mud and water plants, providing a living environment for unique fauna.

References

Aragón Vaca, L.E. (1992) Investigación de ciencias sociales y programas de studio en desarrollo regional en la Amazonía Brasileña: Experiencias, problemas y alternativas, in *América Latina Local y Regional: Memorias de II Simposio de la Universidad de Varsovia sobre América Latina, Varsovia 16–21 de Septiembre de 1991*, Vol. 2. Varsovia: CESLA, Universidad de Varsovia.

Bailey, M. (2013, 26 June) Medically Useless, It's Still Worth more than Gold. *The Art Newspaper* 247. Retrieved 12 December 2016 from http://old.theartnewspaper.com/articles/Thieves-target-rhino-horn/29764.

Bamako Convention (1998) United Nations Environment Programme. Retrieved 12 November 2016 from www.unep.org/delc/BamakoConvention.

Barbosa Magalhães, S. and del Moral Hernandez, F. (2009) Experts Panel Assesses Belo Monte Dam Viability. *Survival International.* October 2009. Retrieved 18 August 2016 from http://assets.survivalinternational.org/documents/266/Experts_Panel_BeloMonte_summary_oct2009.pdf.

Basel Convention (1995) Amendment to the Basel Convention on the Control of Transboundary Movements of Hazardous Wastes and their Disposal. Retrieved 10 November 2016 from www.basel.int/Countries/StatusofRatifications/BanAmendment/tabid/1344/Default.aspx.

Basel Convention (2011) *Where are WEee in Africa?: Findings from The Basel Convention E-Waste Africa Programme.* Retrieved 10 November 2016 from www.basel.int/Portals/4/Basel%20Convention/docs/pub/WhereAreWeeInAfrica_ExecSummary_en.pdf.

Batker, D. and Schmidt, R. (2015) *Environmental and Social Benchmarking Analysis of the Nautilus Minerals Inc. Solwara 1 Project.* Tacoma: Earth Economics. Retrieved 9 February 2016 from http://nus.live.irmau.com/irm/content/pdf/eartheconomics-reports/earth-economics-may-2015.pdf.

Bauman, Z. (2007) *Szanse etyki w zglobalizowanym świecie.* Kraków: Wydawnictwo Znak.

Biosecurity (n.d.) Ministry of Primary Industries, New Zealand. Retrieved 10 November 2016 from www.biosecurity.govt.nz.

Brigden, K., Labunska, I., Santillo, D. and Allsopp, M. (2005) *Recycling of Electronic Wastes in China & India: Workplace & Environmental Contamination.* Exeter: Greenpeace. Retrieved 10 November 2016 from www.greenpeace.org/international/Page Files/25134/recycling-of-electronic-waste.pdf.

BSR (Business for Social Responsibility) (2010) Conflict Minerals and the Democratic Republic of Congo Responsible Action in Supply Chains, Government Engagement and Capacity Building. Retrieved 19 August 2016 from www.bsr.org/reports/BSR_Conflict_Minerals_and_the_DRC.pdf.

Butler Rh. A. (2009) *Peatlands Conversion for Oil Palm a 'Monumental Mistake' for Indonesia's Long-Term Prosperity, Sustainability.* Retrieved 25 August 2016 from https://news.mongabay.com/2009/06/peatlands-conversion-for-oil-palm-a-monumental-mistake-for-indonesias-long-term-prosperity-sustainability/.

Camargo, M., Giarrizzo, T. and Isaac, V. (2004) Review of the Geographic Distribution of Fish Fauna of the Xingu River Basin, Brazil. *Ecotropica* 10: 123–147.

Convention on Biological Diversity (2009) What are Invasive Alien Species?. Retrieved 15 November 2016 from www.cbd.int/idb/2009/about/what/.

Depowski, S., Kotliński, R., Ruhle, E. and Szamałek, K. (1998) *Surowce mineralne mórz i oceanów.* Warsaw: Wydawnictwo Naukowe Scholar.

The Economist (2010, 24 June) The Campaign Against Palm Oil: The Other Oil Spill. *The Economist.* Retrieved 25 August 2016 from www.economist.com/node/16423833?story_id=16423833.

EEA (2016) Protecting Marine Life in Europe's Seas. Retrieved 29 December 2016 from www.eea.europa.eu/highlights/protecting-marine-life-in-europe2019s-seas-1.

Elizondo, G. (2012, 20 January) Q&A: Battles over Brazil's Biggest Dam. *Aljazeera.* Retrieved 19 August 2016 from www.aljazeera.com/indepth/features/2012/01/2012 12011183675441.html.

Ellis K. (2013) *Tackling the Demand for Rhino Horn.* Save the Rhino. Retrieved 15 November 2016 from www.savetherhino.org/rhino_info/thorny_issues/tackling_the_demand_for_rhino_horn.

Enquete Commission (1990) *Protecting the Tropical Forests: A High-Priority International Task. 2nd Report of the Enquete Commission 'Preventive Measures to Protect the Earth's Atmosphere'*, German Bundestag (ed.) ss. 968. Bonn: German Bundestag.

European Commission (2014) Communication from the Commission to the European Parliament. Brussels: European Commission.

Fearnside, Ph.M. (2009) As hidrelétricas de Belo Monte e Altamira (Babaquara) como fontes de gases de efeito estufa. *Novos Cadernos NAEA* 12(2): 5–56. Instituto Nacional de Pesquisas da Amazônia (INPA).

Galapagos Conservancy (2016a) Invasive Snail Detecting Dogs. Retrieved 10 November 2016 from www.galapagos.org/conservation/conservation/project-areas/ecosystem-restoration/invasive-snail-detection-dogs/.

Galapagos Conservancy (2016b) Post-Rat Eradication and Monitoring on Pinzón. Retrieved 10 November 2016 from www.galapagos.org/conservation/conservation/project-areas/ecosystem-restoration/rat-eradication.

Gallup International (2015, 13 April) Losing Our Religion?: Two Thirds of People Still Claim to be Religious. Retrieved 10 November 2016 from www.wingia.com/en/news/losing_our_religion_two_thirds_of_people_still_claim_to_be_religious/290/.

Gao, Y., Stoner, K.J., Lee, A.T.L. and Clark, S.G. (2016) Rhino Horn Trade in China: An Analysis of the Art and Antiques Market. *Biological Conservation* 201: 343–347.

Garrigan, K. (2014) Ivory: From Bush to Market. Retrieved 29 October 2016 from www. awf.org/blog/ivory-bush-market.

Gatimu, S. (2016) Congo-Kinshasa: The True Cost of Mineral Smuggling in the DRC. *All Africa: Institute for Security Studies.* Retrieved 4 January 2017 from http://allafrica. com/stories/201601120281.html.

Global Fishing Watch (n.d.) Retrieved 16 August 2016 from http://globalfishingwatch. org/.

Gormley, A.M., Slooten, E., Dawson, S., Barker, R.J., Rayment, W., du Fresne, S. and Bräger, S. (2012) First Evidence that Marine Protected Areas can Work for Marine Mammals. *Journal of Applied Ecology* 49(2): 474–480. Retrieved 28 December 2016 from www.otago.ac.nz/.

Gouby, M. (2015, 16 March) Democratic Republic of Congo Wants to Open Up Virunga National Park to Oil Exploration. *The Guardian.* Retrieved 28 December 2016 from www.theguardian.com/environment/2015/mar/16/democratic-republic-of-congo-wants-to-explore-for-oil-in-virunga-national-park.

Góralczyk, B. (2016) *Morze Południowochińskie – początek Wielkiej Gry.* Retrieved 16 August 2016 from http://wiadomosci.wp.pl/kat,1027191,title,Morze-Poludniowochinskie-poczatek-Wielkiej-Gry,wid,18439476,wiadomosc.html.

Górecka, A. (2011) Produkcja oleju palmowego a odpowiedzialność z naturalne środowisko. *Technika – Technologia* 65. Retrieved 23 August 2016 from http:// przemyslspozywczy.eu/wp/wp-content/uploads/2011/06/OLEJ.pdf.

Grid Arendal (2005) Who Gets the Trash?. Retrieved 28 May 2017 from www.grida.no/s earch?query=who+gets+the+trash.

Guilford, G. (2013) Why Does a Rhino Horn Cost $300,000?: Because Vietnam Thinks It Cures Cancer and Hangovers. Retrieved 8 December 2016 from www.theatlantic.com/ business/archive/2013/05/why-does-a-rhino-horn-cost-300-000-because-vietnam-thinks-it-cures-cancer-and-hangovers/275881/.

Haddad, N.M. *et al.* (2015) Habitat Fragmentation and its Lasting Impact on Earth's Eco-systems. *Science Advances* 1(2).

Hair, C.A., Pickering, T.D. and Mills, D.J. (2012) Asia-Pacific Tropical Sea Cucumber Aquaculture. Proceedings of an International Symposium held in Noumea, New Caledo-nia, 15–17 February 2011. *ACIAR Proceedings* No. 136, Canberra: Australian Centre for International Agricultural Research. Retrieved 29 October 2016 from http://aciar.gov.au/ files/node/14423/pr136_asia_pacific_tropical_sea_cucumber_aquacu_10936.pdf.

Heinrich, S., Wittmann, T.A., Prowse, T.A.A., Ross, J.V., Delean, S., Shepherd, C.R. and Casseya, P. (2016) Where did all the Pangolins Go? International CITES Trade in Pan-golin Species. *Ecology and Conservation* 8: 241–253. Retrieved 4 January 2017 from www.sciencedirect.com/science/article/pii/S2351989416300798.

Hobbs, J.J. (2004) Problems in the Harvest of Edible Birds' Nests in Sarawak and Sabah, Malaysian Borneo. *Biodiversity and Conservation* 13: 2209–2226.

International Rhino Foundation (n.d.a) Retrieved 15 November 2016 from http://rhinos. org.

International Rhino Foundation (n.d.b) The World's Rhinos are in Crisis. Retrieved 15 November 2016 from http://rhinos.org/the-crisis/.

Kakapo Recovery (n.d.) The Department of Conservation, New Zealand. Retrieved 10 November 2016 from www.kakaporecovery.org.nz.

Komisja Europejska [European Commission] (2014) *Komunikat Komisji do Parlamentu Europejskiego [Communication from the Commission to the European Parliament], Rady, Europejskiego Komitetu Ekonomiczno-Społecznego i Komitetu Regionów w*

168　*Anna Dudek* et al.

sprawie przeglądu wykazu surowców krytycznych dla UE i wdrażania inicjatywy na rzecz surowców. 26 May. Brussels: Komisja Europejska. Retrieved 8 December 2016 from http://eur-lex.europa.eu/legal-content/PL/TXT/?uri=CELEX%3A52014DC0297.

Krukowska, M. (2016) Wirunga, raj przeklęty. *Tygodnik Powszechny* 47: 40–43.

Langenheim, J. (2015, 15 July) Preventing Ecocide in South China Sea. *The Guardian.* Retrieved 19 December 2016 from www.theguardian.com/environment/the-coral-triangle/2015/jul/15/preventing-ecocide-in-south-china-sea.

Makowski, J. (1999) *Zmiany zasięgu wilgotnych lasów równikowych: Przyczyny i konsekwencje.* Warsaw: University of Warsaw.

Milliken, T. (2014) *Illegal Trade in Ivory and Rhino Horn: An Assessment Report to Improve Law Enforcement Under the Wildlife TRAPS Project.* USAID, TRAFFIC.

Milman, O. (2012, 6 August) Papua New Guinea's Seabed to be Mined for Gold and Copper. *The Guardian.* Retrieved 9 February 2016 from www.theguardian.com/environment/2012/aug/06/papua-new-guinea-deep-sea-mining.

Moskvitch, K. (2012) Unused E-Waste Discarded in China Raises Questions. *BBC.* Retrieved 10 November 2016 from www.bbc.com/news/technology-17782718.

Nahigyan, P. (2014, 10 December) 5 Gyres Publishes First Global Estimate Ocean Plastic Pollution. *Planet Experts.* Retrieved 12 December 2016 from www.planetexperts.com/5-gyres-publishes-first-global-estimate-ocean-plastic-pollution/.

The Nature Conservancy (2016) *The Threat of Invasive Species Disrupting the Natural Balance.* Retrieved 12 November 2016 from www.nature.org/ourinitiatives/habitats/forests/explore/the-threat-of-invasive-species.xml.

Nellemann, C., Redmond, I. and Refisch, J. (eds) (2010) *The Last Stand of the Gorilla – Environmental Crime and Conflict in the Congo Basin.* A Rapid Response Assessment. United Nations Environment Programme, GRID-Arendal, Norway: Printed by Birkeland Trykkeri AS. Retrieved 10 November 2016 from www.grida.no.

NOAA (2011) *Ocean Pollution.* Retrieved 12 December 2016 from www.noaa.gov/resource-collections/ocean-pollution.

NOAA (2008) *Currents.* Retrieved 21 May 2017 from http://oceanservice.noaa.gov/education/kits/currents/05currents3.html.

OECD (2012) *Illegal Trade in Environmentally Sensitive Goods*, OECD Trade Policy Studies. Paris: OECD Publishing.

Okurowski, T. (2015) Gdzie złowić ryby?: Mapa światowych połowów. Retrieved 16 August 2016 from www.lokalizacja.info/pl/transport/mapy/gdzie-zlowic-ryby-mapa-swiatowych-polowow.html#.V7rPbDUcXfc.

Perrings, C. (2005) *The Socioeconomic Links Between Invasive Alien Species and Poverty, Report to the Global Invasive Species Program, Global Invasive Species Program.* New York: University of New York.

Phippen, J.W. (2016, 24 May) The Rhino-Horn Trade Returns to South Africa. *The Atlantic.* Retrieved 8 December 2016 from www.theatlantic.com/international/archive/2016/05/south-africa-rhino-horn/484001/.

Prokurat, S. (2013) Palm Oil – Strategic Source of Renewable Energy in Indonesia and Malaysia. *Journal of Modern Science* March: 425–443. Retrieved 23 August 2016 from http://wsge.edu.pl/files/JOMS/3-18-2013/Sergiusz_Prokurat_JoMS_3-18-2013_joms3.pdf.

Protect Planet Ocean (2016) *Global Facts about MPA and Marine Reserves.* Retrieved 21 December 2016 from www.protectplanetocean.org/collections/introduction/introbox/globalmpas/introduction-item.html.

Pullin, A.S. (2004) *Biologiczne podstawy ochrony przyrody*. Warsaw: Wydawnictwo Naukowe PWN.

Radziejewska, T. and Gruszka, P. (2005) Morskie Obszary Chronione – droga do zachowania bioróżnorodności mórz. *Problemy Ekologii* 5: 243–252. Retrieved 25 August 2016 from http://yadda.icm.edu.pl/yadda/element/bwmetal.element.baztech-article-BAR0-0011-0053.

Richey, J.E., Melack, J.M., Aufdenkampe, A.K., Ballester, V.M. and Hess, L.L. (2002) Outgassing from Amazonian Rivers and Wetlands as a Large Tropical Source of Atmospheric CO_2. *Nature* 416: 617–620. Retrieved 19 August 2016 from www.nature.com/nature/journal/v416/n6881/abs/416617a.html.

Rosenbaum, H. and Grey, F. (2015) *Accountability Zero: A Critique of Nautilus Minerals Environmental and Social Benchmarking Analysis of the Solwara 1 Project*. Retrieved 9 February 2016 from www.earthworksaction.org/files/publications/REPORT-AccountabilityZERO.pdf.

Save the Rhino (n.d.) Retrieved 15 November 2016 from www.savetherhino.org.

Sekinchan (n.d.) Retrieved 29 October 2016 from www.sekinchan.org/swiftlet-farming-sekinchan-selangor-malaysia.html.

Sivarnee (2013, 27 June) Thieves Target Rhino Horn, Medically Useless, it's Still Worth More Than Gold. *CNN iReport*. Retrieved 12 December 2016 from http://ireport.cnn.com/docs/DOC-996130.

Skinner, N. (2014, 19 August) African Elephant Numbers Collapsing. *Nature News*. Retrieved 28 October 2016 from www.nature.com/news/african-elephant-numbers-collapsing-1.15732.

Talley, I. (2010, 30 April) Experts: Oil May Be Leaking at Rate of 25,000 Barrels a Day in Gulf, *The Wall Street Journal*. Retrieved 20 January 2016 from www.wsj.com/articles/SB10001424052748703871904575216382160623498.

Thorburn, C. (2014) Bird's Nest Boom. *Inside Indonesia* 116. Retrieved 8 December 2016 from www.insideindonesia.org/bird-s-nest-boom.

Toral-Granda, V. (2008) Galapagos Islands: A Hotspot of Sea Cucumber Fisheries in Central and South America, in V. Toral-Granda, A. Lovatelli and M. Vasconcellos (eds), *Sea Cucumbers: A Global Review of Fisheries and Trade, FAO Fisheries and Aquaculture Technical Paper*, No. 516, Rome: FAO: 231–253.

Torode, G. (2015, 25 June) Insight – 'Paving Paradise': Scientists Alarmed over China Island Building in Disputed Sea. *Reuters*. Retrieved 19 December 2016 from http://uk.reuters.com/article/uk-southchinasea-china-environment-insig-idUKKBN0P50UD20150625.

UNEP, CITES, IUCN, TRAFFIC (2013) Elephants in the Dust – The African Elephant Crisis. A Rapid Response Assessment. United Nations Environment Programme, GRID-Arendal.

The World Factbook (2016) Retrieved 3 January 2016 from www.cia.gov.

World Heritage Encyclopedia (n.d.) *Agbogbloshie*. Retrieved 12 December 2016 from http://self.gutenberg.org/articles/eng/Agbogbloshie.

WWF (2016) In-Depth: Bringing Back Tigers. *World Wildlife*. Retrieved 29 October 2016 from www.worldwildlife.org/magazine/issues/winter-2016/articles/bringing-back-tigers.

8 Arcs of crises, zones of peace?

The geography of wars, conflicts and terrorism in the twenty-first century

Marek Madej

Introduction: basic definitions

Terms relating to the use of force in international relations (or more broadly in politics as a whole) are undoubtedly basic to disciplines studying these relations, since force is one of the central issues in the interaction of states. At the same time, however, how to precisely define these terms remains a challenge. Although some common constitutive elements can easily be identified in phenomena such as armed conflict (e.g. opposing positions between two or more parties resolved by the use of force) or terrorism (e.g. the use of violence in order to create a psychological effect, namely, fear), the sheer number of research approaches and interpretive methods relating to these issues makes it difficult to find terms to describe them which would be widely understood in a uniform and unambiguous manner. This chapter does not aim to settle such disputes, nor does it propose a comprehensive conceptual framework by which to analyse phenomena associated with the use of force and organized violence in international relations (such as can be found in, among others, Shaw 2009: 97–106). Rather, the main intention here is to describe the geographical distribution of the various forms of force used in the rivalry between actors involved in some way in international or transnational relations, or which have an influence on these relations as a consequence of confrontation between states. Therefore, a 'pragmatic' approach is taken to issues of terminology, interpreting the basic terms used in this sphere in line with the most widely accepted views in the literature. Given that subsequent sections of the chapter refer to data on organized violence held in selected databases – especially that of the Uppsala Conflict Data Program (UCDP) at the University of Uppsala in relation to armed conflict, and, in the context of terrorism, the Global Terrorism Database (GTD) maintained by the National Consortium for the Study of Terrorism and Responses to Terrorism, Department of Homeland Defense Center of Excellence, at the University of Maryland – the interpretation of basic concepts used in these databases is followed somewhat closely here.

The most general term in this context remains the concept of organized violence itself, which functions as an *umbrella term* designating all forms of the use of force which are organized to a degree allowing us to speak of a continuity of action targeted towards a specific purpose, and undertaken by an identifiable entity. In general, as exemplified in UCDP reports, organized violence is divided into three basic categories: *state-based conflicts* (which traditionally have been identified with *armed conflicts*), *non-state conflicts* and *one-sided violence*.

The first of these concepts may be called the most traditional, as it relates to the activities of states and is understood as a contested incompatibility relating to a government and/or territory where the use of armed force between two parties, of which at least one is the government of a state, results in at least twenty-five battle-related deaths in one calendar year (SIPRI 2012: 81). The key components of this definition, which in combination reflect the essential character of state-based conflict, are: incompatible positions between two parties; at least one of which is a national government; and whose mutual use of continuous force reaches a given intensity as measured by the number of direct victims of the conflict (i.e. civilian and military fatalities which result directly from combat operations). Such conflicts can be international in character (when they take place between two subjects of international law) or non-international (an internal dispute between a government and a sub-state entity on territory under the authority of the former), or transnational, if a rival extra-state entity is active from outside the borders of a country with which it is in conflict. In addition, such conflicts can take place over who should be in power, or how that power should be exercised, but without calling into question the validity of current political borders – these are known as *conflicts over government*. In addition, they may relate to the sovereign rights of a particular entity (state or non-state) to a given territory, which typically involves a change of borders, or the emergence of autonomous structures in areas, which remain within the existing state (so-called *conflict over territory*). Finally, when in a given calendar year the number of battle-related deaths in a conflict exceeds 1,000, it is automatically classified as a war or major armed conflict (UCDP n.d.).

The second basic category of organized violence – non-state conflict – denotes the use of armed force between two organized armed groups, neither of which is the government of a state, resulting in at least twenty-five battle-related deaths in a year (SIPRI 2012: 81). Due to the fact that today such conflicts outnumber all others, and that the intensity of the violence they entail can at least in some cases reach a level similar to that which characterizes conflicts involving governments, it is reasonable to recognize this category of organized violence as one of the forms of armed conflict. Therefore, this chapter departs from the traditional approach of identifying armed conflicts exclusively with those involving at least one state actor, and treats armed conflict as a generic concept, with two basic subcategories: *state-based conflicts* and *non-state conflicts*.

One-sided violence, the third category of organized violence, is defined as the use of armed forces by the government of a state or by a formally organized group against civilians which results in at least twenty-five deaths in a year (SIPRI 2012: 81). However, as this form of organized violence does not involve the mutual use of force – no armed confrontation between two parties takes place – it is not discussed further in this chapter.

Terrorism is a specific form of organized violence. It is not a category completely disconnected from the various forms of armed conflict already mentioned, as in the course of these, especially in irregular actions by the warring
parties, terrorist acts may be carried out. However, the term also refers to acts of
violence which go beyond the framework of armed conflict or are incompatible
with its essential elements, such as attacks during peacetime. Undoubtedly, the
concept of terrorism itself eludes precise definition due to both the complexity of
the phenomenon, and – above all – the negative moral and legal connotations it
carries (designating an event by this term is tantamount to judging it a crime and
an act worthy of condemnation) (Hoffman 2006: 1–43; Laquer 2003: 232–239).
Nevertheless, without seeking to resolve all the uncertainties associated with the
term, terrorism can be understood as violence or its threat used by non-state
actors in order to advance a political programme by arousing fear in a group of
people larger than those directly attacked. Through the pressure created by terrorism, its perpetrators hope to force governmental compromise, or even the
overthrow of the entire political order (Madej 2007: 134). A similar approach,
although less rigorous with regard to the political nature of terrorism, is taken by
the authors of the Global Terrorism Database, according to whom an act of terrorism must meet three criteria: it is aimed at attaining a political, economic,
religious or social goal; there is evidence of an intention to coerce, intimidate or
convey some other message to a larger audience (or audiences) than the
immediate victims; it is outside the context of legitimate warfare activities and
outside the parameters permitted by international humanitarian law (GTD n.d.b).

Geography of conflicts: African heart of darkness, 'hot' Asia and the Middle East, 'oases' of calm in Europe and the Western hemisphere[1]

Data gathered by researchers at the University of Uppsala reveals clear geographical differentiation in the distribution of conflicts in the world of the
twenty-first century (Table 8.1). In other words, looked at in terms of quantity –
that is, purely the number of conflicts – we can speak about the existence of specific *zones of peace* and *zones of crisis*. From this perspective, the region most
affected by war is Africa, where 247 of the world's 382 armed conflicts in
2001–2013 took place. It is important to note, however, that this high figure is

Table 8.1 The number of armed conflicts in 2001–2013 – regional distribution

	Europe	Western hemisphere	Middle East	Asia and Oceania	Africa	World
Non-state conflicts	0	23	20	44	210	297
State-based conflicts	5	4	9	30	37	85
Total conflicts	5	27	29	74	247	382

Sources: own elaboration based on: Petersson and Themnér (2012), and SIPRI (2012).

dominated by non-state conflicts. The much higher frequency of these in Africa compared to other regions is decisive for why Africa accounts for almost two-thirds (65 per cent) of all conflicts in the world (and 71 per cent of non-state conflicts). However, if we focus exclusively on conflicts involving states (or rather governments) on at least one side of the conflict, then the picture is not as clear-cut. The number of such conflicts in Africa, though still the highest in the world, does not differ so much from those recorded in the twenty-first century in Asia (including Oceania). These are, respectively, thirty-seven (43.5 per cent of all such conflicts) and thirty (35 per cent). However, Asia and Oceania account for significantly fewer conflicts between non-state actors – about 15 per cent (forty-four out of 297 conflicts).

The frequency (in this case low) of non-state conflicts is also the decisive reason why certain areas can be considered 'zones of (relative) peace'. The statistics indicate that the primary example of such an area is Europe, where there are no non-state conflicts at all, and only relatively few (just five in the period 2001–2013) conflicts involving governments.[2] Interestingly, the number of the latter is even slightly smaller for the Americas in the twenty-first century (four in the period 2001–2013, which includes the confrontation between the USA and Al-Qaeda – along with its associated terrorist organizations: a conflict which, excepting its outbreak with the attacks of 11 September 2001, has been fought essentially outside the American continents). However, what distinguishes the Western hemisphere from Europe, and determines that this region, although relatively calm, is an area with a greater intensity of conflicts, is precisely the more frequent occurrence of armed struggle between non-state actors (twenty-three cases). In this context, the situation in the Western hemisphere is unique, and also somewhat controversial, as with just a few exceptions the non-state conflicts seen there in the twenty-first century have taken place between criminal organizations and thus have been primarily motivated by economic factors (or, at the very least, we can say that in these conflicts no explicitly political demands have been raised, or claimed as justification for the use of violence). If we accept that the apparently non-political nature of these clashes 'disqualifies' them as armed conflicts, the situation in the Americas can be considered little different from that prevailing in Europe, and both regions undoubtedly deserve to be called 'zones of peace'. However, if we take into account the intensity of the violence in confrontations between criminal groups in Latin America (which is significantly higher, especially in the case of Mexico, than the average for politically motivated non-state conflicts around the world), as well as the increasing frequency of such clashes in recent years (roughly the last decade) with the numbers of 'battle-related deaths' exceeding the casualty threshold for armed conflict, it can be argued that recognizing this type of violence as a form of armed conflict is justified. This in turn would mean a noticeable difference between the degree of 'peacefulness' in Europe compared to the Americas.

The Middle East region is a special case. It is the smallest region denoted in the SIPRI and University of Uppsala statistics (it has just fifteen countries, while the biggest geostrategic region – Africa – has fifty-three), and this is of obvious significance when we look at the number of armed conflicts recorded for the region in absolute terms. Despite its size, the total number of armed conflicts in the Middle East, is second only, though by far, to the regions of Africa and Asia-Pacific. This relatively high frequency of armed conflicts in the Middle East is emphasized when we compare their number with the number of countries that make up the region – the fifteen countries accounted for twenty-nine conflicts between 2001 and 2013 (and in the years following this number has dramatically increased due to the deterioration of the situation in Syria and Iraq), which means an average of close to two (1.9) conflicts per country. Although this is still much less than in the case of Africa (4.7), it is more than in Asia (1.7), the Western hemisphere (0.8) and Europe (0.1).

Unsurprisingly, there is also a clear differentiation in the frequency of conflicts within regions, especially the largest ones. On the African continent in the twenty-first century a particularly large number of conflicts have been recorded in the area known as the Horn of Africa, due to the relatively high incidence of various, mainly non-state, armed conflicts in Ethiopia and Somalia (thirty-two and thirty-nine conflicts respectively). However, it is good to remember that in addition to the number of conflicts, other key elements are their length and intensity – in Ethiopia of the thirty-two conflicts between 2001 and 2013, twenty-six lasted no longer than a year, and none exceeded the intensity threshold for war (1,000 battle-related deaths in any calendar year of the duration of a conflict). Another sub-regional 'area of particular concern' is West Africa, mainly due to the highly unstable situation in Nigeria (thirty-eight armed conflicts in the twenty-first century). In recent years, a noticeable increase has also been seen in the frequency of conflicts in the Sahel region and North Africa, which is undoubtedly related to the Arab Spring and the spread of radical Islam. Against this background, we can even speak of the decline in significance of the 'war zones' of Central Africa and the Great Lakes region, especially in relation to at least the first decade of the post-Cold War period. At that time, on the territory of the present-day Democratic Republic of the Congo (formerly Zaire), an exceptionally large number of armed conflicts took place both involving state entities and between various non-state communities (twenty-one conflicts in year 2001–2013, ten of which occurred in 2001–2003).

The least stable part of Asia is South Asia, due to the large number of conflicts recorded in Pakistan (twenty-four, of which as many as four were new in 2013), India (sixteen internal conflicts and one international with Pakistan) and Afghanistan (seven conflicts). There have, however, been no open armed conflicts in East Asia, despite continued political tensions and the presence of a number of significant military powers (China, Japan the two Koreas).

A clear concentration of violence can also be observed in specific parts of the Americas and in one country in particular – Mexico (sixteen conflicts, exclusively non-state). This, of course, is due to the intensity of the rivalry between the various drug cartels in the country for control of criminal activities in particular areas, and especially smuggling routes to the United States. In addition, the growth of violence in Mexico has resulted in a certain degree of 'conflict contagion' (Buhaug and Gleditsch 2008: 215–233) and spreading it to neighbouring countries, seen especially in the intensified rivalry between criminal organizations (the *maras*) in Central America and in Guatemala in particular. In addition to Mexico, the other country of the region with a higher than average for the Americas frequency of armed conflicts in the twenty-first century is Columbia. There, however, armed conflict has not been confined to the activity of criminal groups, as conflicts of a political character – at least declaratively – have been prominent (see Box 8.1).

Box 8.1   When crime turns into armed conflict: the cartel wars in Mexico

The crisis in Mexico relating to what are known as *Transnational Criminal Organizations* (TCOs) can be traced back to the 1980s when local drug smuggling gangs, who up to that time had been fully controlled by Colombian cartels, began to assert their independence. Nevertheless, it was only in the first years of the twenty-first century that the scale of their violence against the population, law enforcement agencies and rival groups took on extreme proportions. To answer this challenge, in 2006, the then president of Mexico, Vicente Fox, deployed regular military units against the TCOs. Despite the initial successes of this strategy – including the capture or elimination of twenty-five out of the thirty-seven most important cartel leaders – it also resulted in the existing criminal structures splitting into smaller groups ruthlessly competing for influence and access to smuggling routes. Consequently, the intensity of the fighting between the various cartels and their clashes with the police grew rapidly, as did the number of victims of *drug related violence* (in 2008 – approximately 5,400 fatalities; in 2009 – 6,900 fatalities; in 2011 –12,000 fatalities; and in 2013 – as many as 17,000 fatalities). The main participants in the conflict were the following cartels: Golfo, Sinaloa, Juarez, Tijuana, Beltran Leyva and La Familia Michoacana, who over time began to focus on strategic alliances – the 'Zetas' and 'Anti-Zetas'. The deteriorating security situation led to a further weakening of the local and federal authorities together with the law enforcement agencies, the spread of corruption and also the emergence of formally illegal citizens' self-defense formations to resist the impunity of the cartels. Other forms of crime (kidnapping and human trafficking) also developed, and the situation in Mexico led to tension with the United States and had a destabilizing effect on neighbouring countries (e.g. violence between gangs – or *maras* – in Guatemala). In recent years, the intensity of the violence in Mexico has lessened somewhat, and in 2016, the government of Peña Nieto, president since 2012, succeeded in capturing Joaquín 'El Chapo' Guzmán, the leader of Mexico's most powerful Sinaloa Cartel; however, the scale of the fighting is still at a level which hitherto has characterized only politically motivated conflicts with no criminal origins. Undoubtedly, Mexico's location as a transit country on the route to the most attractive drug market in the world – the USA – has played a considerable role in this. It is also possible that the case of Mexico indicates the direction of the future evolution of armed conflict.

Europe is an interesting case, as its generally low number of recorded conflicts and absence of rivalry between non-state actors of an intensity exceeding the armed conflict threshold allows us to describe the vast majority of the region as being free of organized violence. The armed conflicts recorded in Europe have occurred only on the region's periphery – the Balkans (the Macedonian–Albanian conflict in 2001), and above all in the post-Soviet area, mainly in the Russian Caucasus (and also the 2014–onwards conflict in eastern Ukraine, which as pointed out above, is a clear outlier in terms of intensity compared to the European average).

Among all the geostrategic regions, the most balanced geographical distribution of armed violence is found in the Middle East. Nevertheless, even there, especially in recent years, we can observe a concentration of violence in two countries: Syria and Iraq (nine and five conflicts respectively, bearing in mind that Syria has been a field of battle only since 2011). However, the frequency of conflicts is also increasing throughout the region, although more slowly than in Iraq and especially Syria which saw no fewer than six new armed conflicts in 2013. The region-wide tendency is evidenced by the fighting erupting in Yemen, Lebanon and Egypt.[3] On the other hand, the countries of the Gulf and the Arabian Peninsula (with the exception of Yemen) can be considered 'calm', that is, free from regular armed conflict.

The geography of terrorism: explosive Asia and Middle East, African slippery slope, not-so-calm Europe and 'America feliz (?)'[4]

The twenty-first century to date has come to be rather widely regarded as an 'era of terrorism'. Undoubtedly, the 11 September 2001 Al-Qaeda attacks on the east coast of the United States, which can be considered to represent the symbolic beginning of the new century, has had a huge impact on the direction of the evolution of the international order. This was not only due to their unprecedented scale and psychological effect (2,997 people were killed in four simultaneous attacks, the largest ever number of victims of a single terrorist operation). Also important was the exceptional character of the response made by the United States, which considered the attacks as an act of aggression justifying self-defense, including in the form of military action, and so, as 'an unintended by-product', dignified terrorism by identifying it as one of the key threats to global security (although it must be said that this happened with the tacit approval of a large section of the international community and the active collaboration of many countries, as evidenced by the widespread participation in the anti-terrorist coalition formed at that time and support for UN Security Council resolutions 1268 and 1273 adopted following the September 11 attacks) (Schrijver 2004: 55–74). Somewhat paradoxically, however, this 'appreciation' of terrorism as a threat to international stability, as well as the decisive reaction to its development, has not actually led to any significant decrease in terrorist activity. On the contrary, the war on terrorism has most likely contributed to its growth in essentially all regions of the world.

In general, the 'geography of terrorism' exhibits certain similarities to the territorial distribution of armed conflicts in the world, but important differences can also be observed (Table 8.2). The region most strongly affected by terrorism in the twenty-first century is not Africa, but – at least in absolute values (the total number of attacks) – Asia (formally together with Oceania, but the latter is of marginal significance in this context). In the Asia region, the period 2001–2013 (analogous to that studied in Chapter 7), witnessed 22,973 terrorist attacks (or events suspected of being such). This was 47.4 per cent of all terror acts worldwide which took place in the same period – 48,516 (for the period 2001–2015, the analogous data is as follows: 33,041 out of 74,402 attacks, that is, 44.4 per cent). Importantly, Asia is the area with the highest number of terrorist attacks in each calendar year, in the period 2001–2015.

The region with the second largest absolute number of recorded acts of terror is the Middle East (14,915 attacks in 2001–2013, that is, 30.7 per cent of the total; for 2001–2015, the figures are 24,441 and 32.7 per cent). Worthy of note is the scale of the change which has occurred in the Middle East with regard to the number of recorded attacks. In light of subsequent events, it comes as a surprise that in the first years of the twenty-first century (2001–2002), the Middle East was the region with the lowest number of terrorist attacks in the world (247 in 2001; and 185 in 2002). However, in 2014 and 2015, it was second only to Asia, and that by just a small margin (see Table 8.2).

Although Africa has suffered the most from armed conflict, especially of the non-state variety, it is third in the world as regards terrorism, even after taking into account terrorist acts recorded for North Africa, a region which is culturally

Table 8.2 The number of terrorist attacks in 2001–2015 – regional distribution

	Africa	Asia and Oceania	Middle East	Western hemisphere	Europe	Total
2001	260	592	247	272	485	1,856
2002	240	442	185	194	212	1,273
2003	131	480	195	156	211	1,173
2004	83	457	402	61	108	1,111
2005	134	770	718	72	181	1,875
2006	232	1,173	995	66	168	2,634
2007	380	1,285	1,210	67	134	3,076
2008	439	2,134	1,391	167	380	4,511
2009	341	2,459	1,244	180	369	4,593
2010	358	2,370	1,340	167	394	4,629
2011	481	2,291	1,517	120	282	4,691
2012	1,012	3,580	1,918	142	338	6,990
2013	1,035	4,940	3,553	184	392	10,104
2001–2013	5,126	22,973	14,915	1,848	3,654	48,516
2014	2,383	5,206	5,076	281	696	13,642
2015	2,067	4,862	4,450	200	665	12,244
2001–2015	8,541	33,041	24,441	2,329	5,015	74,402

Source: own elaboration based on: Global Terrorism Database, www.start.umd.edu/gtd.

and politically close to the countries of the Middle East. In 2001–2013, across the continent 5,126 terrorist attacks were recorded, that is, 10.6 per cent of the worldwide total (8,541 and 11.5 per cent in 2001–2015). Moreover, in 2003, Africa was the region least subject to terrorism (131 attacks, although still 11.2 per cent of all attacks). By comparison, eleven armed conflicts involving governments were then taking place (34.4 per cent of the total for this category in 2003) (Pettersson and Wallensteen 2015: 529). In recent years, however, the number of terror attacks in Africa has dramatically increased (from 481 in 2011 to over 1,000 in 2012 and 2013, with as many as 2,383 attacks in 2014).

The Western hemisphere and Europe are relatively calm in terms of terrorism just as they are for armed conflict. However, the two regions do differ as regards the incidence of terror attacks: Europe is much less safe than the Western hemisphere (in the former, 3,654 attacks for the period 2001–2013, representing 7.5 per cent of the total; and in the latter – 1,848 attacks or 3.8 per cent of the total). Furthermore, this disproportion between the Americas and Europe, at least in relation to the number of recorded attacks, is greater than for armed conflicts. It is symptomatic that Europe, which in the twenty-first century has consistently – at least up to 2014 (the year the conflicts in the Ukraine started) – experienced the fewest armed conflicts in the world, was, in no calendar year of the same period, the region with the lowest number of terrorist attacks.

Of course, the picture presented above changes when we take into account the size of the regions themselves and the number of countries they represent. Then, clearly, the part of the world most affected by the problem of terrorism is the Middle East – where, between 2001 and 2013, there were nearly 1,000 attacks (994) on average per country – and when that period is extended by two years (2001–2015), the figure is even higher – 1,629. For Asia, these proportions are somewhat different – 522 (2001–2013) and 751 (2001–15), for Africa the respective figures are ninety-six and 161, and Europe – seventy-seven and 102. The Western hemisphere has the best results in this context – an average of fifty-three terror attacks per country in the period 2001–2013, and sixty-seven in the period 2001–2015.

Uneven distribution within regions is even more noticeable in the case of terrorism than for armed conflict. Terrorist attacks are generally concentrated in only a few countries of a given region. To take Asia as an example, 75 per cent of the region's terror attacks in the twenty-first century have taken place in only three (admittedly vast and populous) countries: Pakistan (6,885 attacks in the period 2001–2013; and 9,741 in 2001–2015), Afghanistan (4,976 in 2001–2013 and 8,303 in 2001–2015 respectively) and India (4,739 in 2001–2013 and 6,316 in 2001–2015). Furthermore, given that in the same period there were also numerous attacks in Nepal (777 – in years 2001–2013; 830 in years 2001–2015), Sri Lanka (655 in 2001–2013 and 680 in 2001–2015 respectively) and Bangladesh (299 in 2001–2013 and 866 in 2001–2015), we can say that South Asia was by far the hardest hit by the problem of terrorism, not only in the region (80 per cent of the attacks in Asia in 2001–2013), but also in the world. At the same time, other countries in the region, even one as large and internally diverse as China, experienced terrorism to just a limited or even negligible degree.

A similar, though not quite as clear concentration of terrorist activity only in some countries can be found in the Middle East, where in Iraq alone there were as many as 11,162 attacks in the period 2001–2013 (16,941 in 2001–2015). Iraq has suffered more than any other country at the hands of terrorists (with approximately 23 per cent of all acts of terror in the world having taken place there). Egypt has also seen a relatively large number of attacks in the twenty-first century (1,069 attacks up to 2016, of which only thirty had occurred up to 2010), as have Turkey (836 up to 2016), Israel and Palestine (1,853 up to 2016, of which 850 occurred in the area of the Palestinian Authority), Yemen (1,761, of which only eighty-four before 2010), and in recent years also Syria due to the outbreak of internal conflict (1,066 attacks up to 2016, but only three before 2010). At the same time, the number of terrorist attacks recorded in the Gulf states (Saudi Arabia, UAE, Oman, Qatar, Kuwait and Bahrain) was relatively low (a total of 289 attacks to the end of 2015, mainly in Saudi Arabia and Bahrain, 139 and 142 respectively; that constitutes 1.2 per cent of attacks in the region).

Also, in the case of Africa, there are certain areas of particularly intensive terrorist activity. One of these is North Africa, where terrorism has been a feature of essentially the entire period since 2001 (1,214 attacks to the end of 2013, but 2,276 in the period 2001–2015). A high number of acts of terror have also been recorded – at least since 2007 – in East Africa, including in particular Somalia (1,101 attacks up to 2014, and 1,819 up to 2016), and – to a lesser degree – Kenya (240 attacks up to 2014 and 400 attacks up to 2016 respectively). In recent years, especially after 2011, the intensity of terrorism has sharply risen in Nigeria (1,361 attacks in the period 2001–2013, and 2,623 in years 2001–2015, but only 476 to the end of 2011) and its neighbours (especially Cameroon – 143 incidents in years 2001–2015 but only twelve up to 2014). This has mainly been due to the activity of Islamic fundamentalists from Boko Haram. Central Africa, including the Democratic Republic of the Congo (393 attacks up to 2016), remains an important but clearly less significant centre of terrorist activity. In this part of Africa, the various inter-ethnic rivalries usually find expression in more traditional forms of organized violence, closer to armed conflict.

In Europe, the area most impacted by terrorist activity since 2001 (but also earlier) is Russia, and especially its Caucasus region (1,476 incidents in the period 2001–2013, and 1,536 in years 2001–2015 – or 40 per cent and 30 per cent respectively of all attacks in Europe). It is difficult to identify any other country in Europe which could be said to constitute a 'centre' of terrorist activity. There have been a negligible number of attacks in the countries of the Visegrad Group (nineteen attacks in the period 2001–2015, and four fatalities), in Scandinavia (seventy-nine incidents, eighty-five fatalities, of which sixty-nine occurred as a result of one attack by 'lone wolf' Anders Breivik on Utøya island in Norway 2011), Benelux (thirty incidents) and the Baltic countries (five incidents). In the Americas, Columbia is clearly a country seriously affected by terrorism, as this is where most of the region's terror attacks have been recorded (1,297 in the period 2001–2013, and 1,602 up to the end of 2015). The United States (222 in the period 2001–2013 and 282 incidents up to the end of 2015) has seen much lower levels of terrorist activity, and no small number of countries in the Western hemisphere are almost entirely free from the threat of terrorism (especially the Caribbean area – just thirteen attacks in 2001–2015 – along with Brazil, Argentina and Uruguay – thirty-six incidents in total to the end of 2015).

Yet another picture of the distribution of terrorist activity in the world can be gained by factoring in the intensity of terrorist attacks as measured by the number of fatalities. In the period 2001–2013, the most bloody attacks – causing 100 deaths or more – took place primarily in the Middle East and Asia, with fifteen such attacks in each region (thirteen in Iraq and three each in Pakistan and Nepal). Seven such incidents took place in Africa (including two in Nigeria), four in the Western hemisphere (of which three were the September 11 attacks, and the fourth was an attack in Colombia in 2002) and two in Europe (exclusively Russia). Interestingly, in the next two years (to the end of 2015), the number of such major acts of terror increased significantly in the Middle East (to twenty-eight, while of the thirteen new attacks seven occurred in Iraq and four in Syria), and Africa (to twenty-four, with eleven of the seventeen new attacks occurring in Nigeria and three in Cameroon). In the same period, two more such attacks were recorded in Europe (Ukraine) and Asia (Afghanistan and Pakistan), while in the Western hemisphere, there were no further major attacks. Taking into account attacks with a smaller, but still a relatively high number of fatalities (between fifty and 100), the highest values were still recorded in the Middle East (fifty-seven such attacks in 2001–2013, and seventy-nine to the end of 2015), Africa (thirty-five in 2001–2013, and seventy-three to the end of 2015, of which thirty-four occurred in Nigeria) and Asia (forty in 2001–2013, and forty-seven to the end of 2015, mainly in Pakistan and Afghanistan). In the same period – 2001–2013 – only five such acts occurred in Europe (this increased to six by the end of 2015 due to the Paris attack of November 2014) and none at all in the Western hemisphere.

Multinational armed interventions: wider Middle East

The period after 2001 has also been marked by increasing numbers of multinational interventions. This has largely been in connection with the intensification of the fight against terrorism (especially following the declaration of the 'Global War on Terror' by the United States after the attacks of 11 September 2001). It has also meant more frequent willingness to address threats to global or regional security, and thus to contribute to greater international stability, by the use of military force on the part of groups of countries – mostly Western, but with the consent (though not necessarily enthusiastic) of a substantial number of the other members of the international community. The multilateral military interventions undertaken in this period, of which Western countries have been the sole or main initiators and executors, have not, however, constituted wars in the traditional sense. For they have been initiated – at least officially – to further the interests and will of the international community as a whole and not only those of the countries directly involved (Rose 2010). At the same time, these interventions go beyond the bounds of traditional peacekeeping – they have the character of enforcement actions, and thus involve a combat (*kinetic*) element, which takes the form of the deployment of military units or periodic air and naval operations.

As would be expected, such interventions have occurred far less often than armed conflicts and terrorist campaigns. Decisions are made to engage in them – at least in theory – only where there are conflict situations which, in the judgement of a larger group of countries (not necessarily limited to those directly intervening) could lead to serious consequences on an international scale, especially the growth of worldwide terrorist activity. Thus we can include in such operations the intervention in Afghanistan that begun in 2001, the operation against Iraq in 2003 and the subsequent stabilization campaign which continued to 2010, the intervention in Libya in 2011 and the campaign by the US-led coalition against ISIS since 2014.[5] Importantly, these operations – once entered into – have proved to be of above average intensity in terms of the scale of the use of force (quantity of troops and equipment) and the number of victims of the fighting. For example, in 2005 during the Iraq intervention, which followed the initial 2003 operation more than 180,000 soldiers were deployed (Iraq Index: 19), and in Afghanistan approximately 140,000 soldiers were involved at the height of the ISAF mission (2010–2011) (Livingston and O'Hanlon 2014: 4–5). It is characteristic, too, that these operations have mainly taken place in the broadly defined Middle East and its environs (Afghanistan and Libya). This concentration of interventions may be evidence of the particular geostrategic value attached by the international community as a whole (and the West in particular) to the Middle East region and the ongoing conflicts there. At the same time, however, the course of all of the above-mentioned international interventions (perhaps with the exception of the anti-ISIS campaign) would seem to show that the effectiveness of the interventions in relation to the declared aim of building international stability and reducing the intensity and frequency of conflicts, as well as – and above all – curbing terrorist activity, has been far from satisfactory. In no country subject to this interventionist activity has there been a lasting end to the fighting, and in the case of Iraq and Libya, intervention has proved to be counterproductive. This has additionally contributed to a sharp increase in terrorist activity not only in the Middle East, but also in neighbouring regions (especially Africa and Europe). It is also possible that the clearly limited effectiveness of the multilateral interventions has contributed to the reduction over time in their international make-up. The Iraq and Afghanistan operations garnered the participation of forty and fifty-one countries respectively. However, although the coalition intervening in Libya was formally comprised of the whole of NATO and some other partner countries, combat operations were conducted by just eight allies (of twenty-eight) and four partners. Attacks on ISIS are carried out by just a dozen or so countries, despite the official anti-ISIS coalition being made up of more than sixty countries (McInnis 2016, 1013) (see Box 8.2).

Box 8.2　Western wars of choice: the interventions in Afghanistan and Iraq

The intervention in Afghanistan (2001–2014). Military action in Afghanistan took place after the 11 September 2001 terrorist attacks in the USA and was sanctioned by UN Security Council Resolution 1368 (2001) as a justified act of self-defense by the USA in response to an armed attack. The aim of the intervention in Afghanistan was to destroy the structures of Al-Qaeda there responsible for the 9/11 attacks and to overthrow the ruling Taliban, who provided support to that organization. *Operation Enduring Freedom* (OEF), conducted mainly in the form of air strikes by the US-led 'coalition of the willing' and supported by the anti-Taliban Afghan opposition (Northern Alliance), was launched on 7 October 2001. Within two months, the Taliban government was overthrown and Al-Qaeda structures in Afghanistan had been significantly weakened. By 22 December, a new Afghan government was sworn in, and thus the international operation entered its second phase, that of stabilization. The goal was the final defeat of the Taliban and Al-Qaeda (thus a continuation of the OEF mission) and the reconstruction of a stable Afghan state (supported by the ISAF – International Security Assistance Force – formed under a UN mandate). However, this did not bring the conflict to an end. The Taliban insurgency, which was initially weak and concentrated in the south of the country, gradually regained its position, aided by the thin deployment of international pro-government forces, a Taliban *safe haven* in northern Pakistan and the increasingly clear ineffectiveness of the new Afghan authorities. The recovery of the Taliban could not be prevented by OEF and ISAF integration (2006) and the gradual but significant increase in ISAF uniformed personnel (up to approximately 140,000 troops in 2010–2011) and of intensity of anti-Taliban combat operations. Therefore, from 2012 onwards, the coalition focused on the development and training of Afghan government forces to prepare them to take independent control of security in the country. The intervention was ended in 2014 leaving Afghanistan politically and economically unstable and in a state of open armed conflict. More than 90,000 people had died during the Afghan operation (including 3,500 international troops) and the intervening countries had incurred more than a trillion dollars in costs.

The intervention in Iraq (2003–2010). The operation *Iraqi Freedom*, also conducted by the US-led Coalition of the Willing, undertaken in connection with allegations (ultimately unproven) that the government of Saddam Hussein in Iraq had developed illegal unconventional weapons and was supporting terrorists. The aim of the mission was to overthrow Saddam Hussein and create conditions for systemic political change in Iraq. Unlike OEF in Afghanistan, the Iraq intervention was not sanctioned by the UN Security Council. Air strikes against Iraq began on 19 March 2003, Baghdad was occupied by the Coalition on 9 April, and on 1 May of that year US President G.W. Bush announced the end of 'major combat operations'. Nevertheless, already in May insurgent and terrorist activities were being directed against the Coalition, conducted not only by supporters of the toppled regime, but also by the Sunni minority and even the Shiite majority which had suffered discrimination under Hussein. The spring of 2004 saw the outbreak of a Sadrist uprising led by the radical Shiite Mahdi Army. In the following year, despite the adoption of a new constitution and the transfer of power to a sovereign

Iraqi government, fighting continued to grow, until in 2006, it acquired the additional feature of Shiite–Sunni sectarian violence. Some improvement in the situation was brought by the *surge* strategy implemented by the USA and its allies in early 2007, under which there was a dramatic increase in the number of combat forces and the intensity of their activities. However, this merely reduced the intensity of the conflict, without ending it. In February 2009, President Barack Obama declared the intention of the USA to end its involvement in Iraq. The main bulk of the intervention forces were withdrawn in August 2010 and the remainder by the autumn of 2011. Iraq, however, continued in a state of civil war, which finally enabled the development on its territory of the governmental structures of a new fundamentalist group – Islamic State.

Both interventions, despite certain differences (for example, as to the degree of their legality and international legitimacy), were characterized by the success of the initial phase in which the aim was to overthrow the existing government, and the failure of the subsequent stabilization phase when the intervention forces had to deal with a determined irregular adversary. Without doubt, the fact of the geographical remoteness of the theatres of operations from the country which was their main initiator – the USA – contributed to the decline in the will to continue the commitment when the missions met with serious obstacles.

Conclusions

If the 'maps' of armed conflict and terrorist activities sketched in the analysis above are 'overlaid' on one another, and then information is added concerning multinational military interventions, we will find that although the various forms of organized violence in the twenty-first century have occurred essentially in each geographical region, there are very significant differences between regions in the intensity of its manifestations and differentiation of its categories. The area most seriously affected by organized acts of violence in the twenty-first century is – hardly surprisingly – the Middle East. However, a similar frequency of conflicts and terrorist acts can also be seen in Asia, and especially South Asia which borders on the Middle East. The African continent is not far behind these areas in terms of the intensity of the violence it has experienced, especially in recent years when there has been a sharp increase in terrorist activity. Especially affected are the northern and central parts of the continent – the belt stretching from the Atlantic to the Indian Ocean to the north of the Gulf of Guinea (Maghreb, Sahel, West Africa and the Horn of Africa). Against this background, the Western hemisphere and Europe may be considered much 'calmer' – especially from a 'macro' perspective in which each region is viewed as a totality. It is worth noting, however, that whereas Europe has experienced frequent manifestations of terrorism, armed conflicts on its territory have only occurred occasionally (and clashes between non-state actors, without the participation of governments, are completely absent). The situation of the Americas is different – it has the lowest level of terrorist activity of any region, but at the same time, it has witnessed a significant number (though still not comparable with Africa and

Asia) of armed conflicts (mainly, however, in just two countries – Mexico and Colombia). All of these considerations seem to argue for the continued validity in the twenty-first century of Zbigniew Brzeziński's description of a 'global arc of crisis' or 'global Balkans', stretching from the northern part of the East African coast of the Indian Ocean through the Arabian Peninsula and other areas of the Middle East to the western part of South Asia (Brzeziński 1997: 123–151). Nevertheless – as correctly noted in, for example, the 2008 French White Paper on Defence and National Security – this area has 'expanded' into the belt of African territories extending to the Gulf of Guinea and the Atlantic coast, encompassing the whole North Africa and the Sahel region, including also their western parts (French White Paper 2008: 41–42).

It should be emphasized, however, that the 'macro' perspective in which each geostrategic region is viewed as a totality, gives an imprecise picture of the subject of this chapter. For both in relation to terrorism and armed conflict, the territorial distribution of these phenomena within individual regions is highly differentiated, with many local 'oases of peace' as well as 'centres of violence', not infrequently – as in the case of the Levant and the Persian Gulf – bordering on one another.

An analysis of the geographical distribution of armed conflicts and acts of terrorism throughout the world is certainly not to be considered a method by which we can unambiguously identify the causes of the outbreak of wars and organized violence in general, either on a global scale or within particular regions. It is true that a map of regions marked by a special intensity of this violence, and especially armed conflict, corresponds in no small degree to the distribution of poverty in the world. However, in relation to this correlation, it is difficult to solve the *reverse causality* dilemma and determine whether the conflicts and violence are the result of the region's poverty, or on the contrary, are its cause (Collier 2009: 126–129). Furthermore, as can be seen from the example of the countries of the Middle East, a high intensity and frequency of conflict, especially terrorism, can be accompanied by relative prosperity (or at least not poverty). And at the same time, some poorer countries experience organized violence to a negligible degree, as for example the countries of Oceania – although many of these are island nations and this naturally limits the number of possible conflicts and reasons for terrorism. Similarly ambiguous and inconclusive as to the construction of any hard generalizations about the reasons for conflicts would be a comparison of the geographical distribution of armed conflicts and terrorism with similar 'maps' of authoritarianism, ethnic and religious diversity, the degree of democracy or state failure. For even if some overlap could be observed between areas particularly strongly affected by conflicts, and regions with large ethnic diversity or a low level of democratization, this would not permit us to draw conclusions regarding a causal relationship between these factors. Thus, the geographical distribution of armed conflicts and other forms of organized violence in the world taken on its own does not provide the answer as to how far geography, in other words the particular location in space of a region affected by conflict or terrorism, determines the occurrence of these phenomena.

Nevertheless, the analysis undertaken here clearly points to a dangerous trend – the gradual expansion of the area characterized by a high intensity of conflicts (the above-mentioned belt stretching from the Atlantic coast of Africa in the northern hemisphere via the Middle East to South Asia), and even a globalization of terrorism, all of which means the increasing occurrence of this type of activity in basically every geostrategic region of the globe. Therefore, although we can still point to both arcs of crises and zones of peace in the world, the latter seem to be in decline.

Notes

1 Quantitative tables presented in this chapter, unless otherwise stated, are based on data contained in Petersson, T. and Themnér L. (eds) *States in Armed Conflict 2011* (Uppsala University: Department of Peace and Conflict Research Report 99, 2012) and *The SIPRI Yearbook: Armaments, Disarmament and International Security*, 2012–2015 editions.

2 Table 8.1 naturally does not include the events of 2014 in Ukraine, but even given the tragedy in Eastern Ukraine (according to the UN's Office for the Coordination of Humanitarian Affairs – OCHA – from the outbreak of the conflict to 15 August 2016, some 9,600 people died in the fighting), Europe remains a relatively 'safe' area, especially as despite the high intensity of the Ukraine conflicts, they have not dramatically increased the numerical quantity of conflicts on the continent. See Office of the UNHCHR (2016).

3 In Yemen, the armed conflict between the government and anti-government groups, has been ongoing since 2009, and since 2010 there has been full-scale fighting between non-state actors; in Lebanon, relatively intense armed clashes have been taking place between Sunnis and Alawites since 2012 (which is no doubt a consequence of the outbreak of fighting in neighbouring Syria as a result of the same underlying causes), and in Egypt, in 2013, fighting erupted between supporters and opponents of ousted President Mursi of the Muslim Brotherhood.

4 The figures for terrorist attacks used here are taken from the Global Terrorism Database (GTD n.d.a), and include questionable cases (i.e. when attackers or their motives are not clearly established).

5 Apart from multilateral actions, some countries have also conducted unilateral interventions in the twenty-first century. However, only a small number of these operations have obtained some form of international legitimacy from the United Nations or other organizations, as serving the interests not only of the intervening party, but also the international community as a whole. Unilateral operations, especially in the case of the USA, relate mainly to the fight against terrorism (limited involvement in the fight against rebels in the Philippines, also Colombia in 2002–2003; and drone attacks in Pakistan, Yemen and Somalia), or the need to restore regional stability – for example, US intervention in Liberia in 2003, and Haiti in 2004–2005, France's Operation Unicorn in the Ivory Coast in 2004 and involvement in the conflicts in Chad, especially in 2006 – or a combination of these elements (the French intervention in Mali in 2012 and the Central African Republic in 2013).

References

Brzeziński, Z. (1997) *The Grand Chessboard*. New York: Basic Books.

Buhaug, H. and Gleditsch, K.S. (2008) 'Contagion or Confusion?: Why Conflicts Cluster in Space. *International Studies Quarterly* 52(2): 215–233.

Collier, P. (2000) Rebellion as a Quasi-Criminal Activity. *The Journal of Conflict Resolution* 44(6): 839–853.

Collier, P. (2009) *Wars, Guns and Votes: Democracy in Dangerous Places*. New York: Harper Perennial.

French White Paper on Defence and National Security (2008). Paris: Odile Jacob.

GTD (n.d.a) Global Terrorism Database. Retrieved 2 November 2016 from www.start. umd.edu/gtd/.

GTD (n.d.b) Global Terrorism Database. Data Collection Methodology. Retrieved 2 November 2016 from www.start.umd.edu/gtd/using-gtd/.

Hoffman, B. (2006) *Inside Terrorism*. New York: Columbia University Press.

Iraq Index (2010) *Iraq Index, Tracking Variables of Reconstruction & Security in Post-Saddam Iraq* Brookings, 30 November. Retrieved 2 November 2016 from www. brookings.edu/wp-content/uploads/2016/07/index20101130.pdf.

Laquer, W. (2003) *No End to War: Terrorism in the Twenty-First Century*. New York: Continuum.

Livingston, I.S. and O'Hanlon, M. (2014) *Afghanistan Index*. Brookings, 14 May. Retrieved 2 November 2016 from www.brookings.edu/wp-content/uploads/2016/07/index20140514.pdf.

Madej, M. (2007) Zagrożenia asymetryczne bezpieczeństwa państw obszaru transatlantyckiego. Warsaw: PISM [Polish Institute of International Affairs].

McInnis, K. (2016) *Coalition Contributions to Countering the Islamic State*, CRS Report R44135, 24 August, Retrieved 2 November 2016 from www.fas.org/sgp/crs/natsec/R44135.pdf.

Office of the UNHCHR (2016) Report on the Human Rights Situation in Ukraine 16 May to 15 August 2016. Retrieved 2 November 2016 from www.ohchr.org/Documents/Countries/UA/Ukraine15thReport.pdf.

Petersson, T. and Themnér, L. (eds) (2012) *States in Armed Conflict 2011*. Uppsala: Uppsala University: Department of Peace and Conflict Research Report 99.

Pettersson, T. and Wallensteen, P. (2015) Armed Conflict 1946–2014. *Journal of Peace Research* 52(4): 536–550.

Rose, G. (2010) *How Wars End: Why We Always Fight the Last Battle: A History of American Intervention from World War I to Afghanistan*. New York: Simon & Schuster.

Schrijver, N. (2004) September 11 and Challenges to International Law, in J. Boulden and T.H. Weiss (eds), *Terrorism and the UN Before and After September 11*. Bloomington, IN: Indiana University Press.

Shaw, M. (2009) Conceptual and Theoretical Frameworks for Organised Violence. *International Journal of Conflict and Violence* 3: 97–106.

SIPRI (2012) *SIPRI Yearbook 2012: Armaments, Disarmament and International Security*. Oxford: Oxford University Press.

UCPD (n.d) Uppsala Conflict Data Program. Retrieved 2 November 2016 from www.pcr. uu.se/research/ucdp.

9 Geographies of twenty-first-century disease

Epidemiological versus demographic transition

Izabella Łęcka

Introduction

Health – a subject so important to us because it relates to each person individually – seems to continually evade our attempts to capture it within any single research framework. A key factor which would allow humanity to discover a hitherto unknown way to avoid disease and achieve longevity has proved elusive. More recently, however, it seems that attention is turning to deficiencies in our actions, rather than in our understanding of the problem. Smith (2016), a behavioural economist, points to human insubordination: modern medicine has much advice for the sick, ranging from a healthy diet, increased physical exercise, avoidance of excess alcohol and other lifestyle suggestions all the way through to pharmacological treatment, but patients do not necessarily follow doctors' orders. Many of today's most pressing public health issues would disappear, according to Smith, if people simply made better choices. However, Burch *et al.* (2016) conclude, based on research in Britain, that the making of rational choices is correlated with socio-economic status, and it is the latter which ultimately determines a person's health. For HIV-infected people in the UK, low socio-economic status is strongly linked to failure to adhere to doctors' recommendations and poor treatment outcomes. This conclusion is supported by data from Poland where the worst state of health is found among patients from villages and small towns of fewer than 5,000 inhabitants (generally characterized by lower incomes and lower education), although their living environment is much healthier than in large cities (Wojtyniak and Goryński 2016).

Those who have control over their lives and have the chance to make rational choices are in a better situation. But what of those who have no choice, who find themselves in a difficult social or economic situation, for whom choices are made by the so-called 'global community' (which often acts for its own benefit without realizing that its actions may not be beneficial for everyone)? Is the health and even survival of such people a mere lottery? There are no simple answers to these questions. Mainly because we still are not in a position to give them. State-of-health indicators, which form the basis of current research, are not good enough to provide a properly nuanced analysis. The basic measures of health used actually relate to its opposites, that is, morbidity, and, most often, mortality by cause of death.[1] Within such a general perspective, negative social or economic phenomena are not always linked to unambiguously negative health effects as reflected in the morbidity and mortality rates. For example, in Spain, during the economic crisis after 2008, there was a decrease in mortality, especially among the poorer sections of society (Regidor *et al.* 2016). Therefore, despite the difficult economic situation, there was a statistical improvement in health. This is unlikely to have been the result of rational choices by Spanish workers, but instead, we may assume that due to the global economic crisis, and, specifically, the unemployment which came in its wake, there was a reduction in workers exposed to physical risk factors in labour intensive sectors.

Does this mean that the situation in all European countries affected by the crisis was as 'good' as in Spain? No, because in Greece, also deeply mired in economic crisis, adverse changes in the mortality rate compared to the years before the crisis quickly became apparent. This varied depending on age, sex and cause of death. In total, there were an estimated 242 deaths per month after the onset of the crisis (Laliotis, Ioannidis and Stavropoulou 2016). A number of these can be attributed to a clear increase in the number of deaths due to adverse events during medical treatment, which in turn correlates with the problem of underinvestment in the Greek health care system as highlighted in the media (McKee and Stuckler 2016).

Interestingly, quality of life surveys among children in Spain between 2001 and 2010 showed increased life satisfaction, while responses by children in Greece indicated that their satisfaction with life had declined over the same period (Figure 9.1).

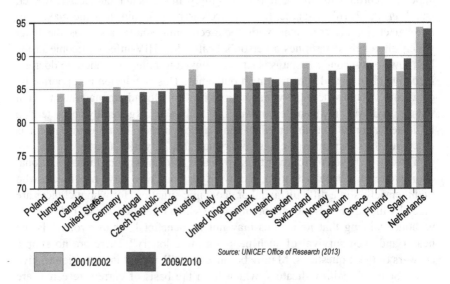

Source: UNICEF Office of Research (2013)

■ 2001/2002   ■ 2009/2010

Figure 9.1 Changes between 2001 and 2010 in the self-reported life satisfaction of children aged eleven, thirteen and fifteen (on the 'Cantril Ladder' scale).

Thus, not only in health research but also studies of well-being, the devil is in the detail and the use of imperfect measures means that there is still too much room left for guesswork.

So what are the factors which it is generally agreed determine our chances for a long and healthy life? Evans, Barer and Marmor (1994) in order to answer the question: 'Why Are Some People Healthy and Others Not?' propose a conceptual framework allowing the merging of research findings from many different disciplines in relation to the role of such diverse factors as: culture, genetic predisposition, biological pathways and social and economic environments.

In sum, the state of human health is determined by the whole of life (not just biological preconditions), including a person's material environment, social relationships, education level, natural environment and medical care (access, quality and institutional framework, including relevant policies). All of these determining factors should be examined on a variety of scales: individual, community, society and finally global. Even with this diversity of approaches, mortality by cause of death continues to be the staple health information statistic. And little is likely to change in this regard.

The concept of health on different scales

Medical geography, health geography, public health, global health – there are so many terms and an ever broader, multidisciplinary conceptualization of the notion of spatial health. At a time when we are trying to think globally, no sooner have we added another aspect to our conception of health, than a further factor immediately appears, which must also be taken into account.

The global scale: Planetary Health is the name of a new discipline which is intended to complement public and global health. Pioneered by Anthony Capon at the School of Public Health in Sydney, it is the broadest conceptualization of spatial health yet developed and it is gaining traction before our very eyes: 2016 saw the formation of The Planetary Health Alliance, an association of universities and non-governmental organizations led by Sam Myers of the Harvard TH Chan School of Public Health in Boston, Massachusetts. In Europe, in the same year, 'The Contribution of Science to Planetary Health' was one of the plenary sessions held during the European Public Health Conference in Vienna, and the Global Health Film Festival in London awarded the first ever 'Planetary Health Film Prize'.

In 2017, a new scientific journal, *The Lancet Planetary Health*, was founded, of which:

> The core idea is the unity of the planet's natural and physical systems. And that the future health of our species demands a different level of investigation – not only of individuals and communities, but also of civilizations, the organization and content of our cultures and societies. The predicaments our species faces are so severe that we must ask far-reaching questions about the capacity of our political, economic, and social institutions to adapt to

and address those predicaments. There is no guarantee that our species, as successful as it is in so many ways, can navigate the challenges we face. Planetary health is concerned with the future survival and flourishing of everything we hold dear.

(Horton 2016: 2462)

The local scale: A characteristic feature of the beginning of the twenty-first century was the remarkable progress achieved in reducing the mortality of mothers and children, which was to some extent the result of measures under-taken to meet the Millennium Development Goals. The decrease in the number of deaths of children under five years old, from an estimated 12.7 million in 1990 to less than 6 million fifteen years later, is an extraordinary achievement in the field of global health care. Unfortunately, this success has not been experi-enced in all the countries of the world equally. It is well known that the mortality of mothers and children varies considerably both between and within individual countries. Furthermore, according to studies conducted in twenty-eight of the countries of sub-Saharan Africa (limited due to imperfect statistical data), the survival chances of mothers and children were not so much influenced by the level of material wealth, but rather 75 per cent determined by other factors such as climate change, malaria and political instability. It was also concluded that local, rather than national and transnational, conditions have the greatest impact on differentiation in mortality. This rule has been confirmed in an increasing number of countries, not only in the developing world but also in highly developed countries. In the USA, in 2006, epidemiological patterns of adult mortality were used to theorize a division of the country into 'eight Ameri-cas' across races, counties and race-counties (Bhutta 2016). So, even though we live in an era of globalization, it is local conditions which directly determine the length and quality of our lives. This may seem trivial as a conclusion of scient-ific study, but in order to reach it, as has repeatedly been the case in the history of public health research, we have had to investigate a successively expanding pool of increasingly global determining factors. A paradox? No, that is how the scientific method works.

The epidemiological transition

The epidemiological transition model has been used for more than forty years, to link relationships between human health (the structure of morbidity) and the eco-nomic development of countries and regions.

The theory of epidemiological transition is derived from the earlier developed theory of demographic transition, which describes the shift from a situation of simultaneously high birth and death rates to the opposite – very low birth rates combined with low death rates. Thomas McKeown initially, and then two years later, in 1971, Abdel R. Omran from the University of Maryland and George Washington University, supplemented the demographic transition theory with the concept of specific causes of death at various stages of demographic change, along with factors which could limit mortality from these illnesses (Omran 1971; Caldwell 2001).

For the next twenty years, the theory developed by McKeown and Omran remained relatively obscure, but then in the 1990s, the changes brought about by the latest wave of globalization led to it gaining in popularity and authority. It had a great impact on the development of public health research, and after it had been expanded to include more detail on factors determining good health maintenance, was designated the health transition theory.

Thomas McKeown argued that health improvements in Western Europe were due to progress in hygiene, together with the improved social situation and living conditions of its population. He also emphasized that increases in the real income of families had resulted in better sanitation and nutrition, which in turn led, as early as the second half of the nineteenth century, to significant decreases in mortality (Link and Phelan 2002; Schreter 2002; Caldwell 2001).

In McKeown's discussion (1975) concerning the fall in mortality in England and Wales in the twentieth century, he divides illnesses, somewhat arbitrarily, into the following groups:

1. infectious diseases transmitted by droplets (through the air), such as tuberculosis, bronchitis, influenza, pneumonia, measles, chickenpox, scarlet fever and diphtheria; 2. infectious diseases transmitted in water and food, such as cholera, diarrhoea, dysentery and typhus; all other illnesses, including infectious and degenerative diseases,[2] as well as a large number of what he calls 'unknown diseases'.

Partly on account of the weakness of this classification, Abdel Omran (1971) developed McKeown's concept by dividing the epidemiological transition into historical periods with different patterns of death and varying mortality rates. According to Omran, the following phases are typical of the epidemiological transition: 1. 'Pestilence and famine', when the number of deaths due to hunger and epidemics increases cyclically; 2. 'Receding pandemics', when the death rate from infectious diseases rapidly decreases and cyclical fluctuations in the number of deaths flatten out); 3. 'degenerative and man-made diseases'.

However, not all countries have followed the classical epidemiological transition scenario as described above. In underdeveloped countries, the tendency of change has been different, and in Western Europe variations in mortality trends and disease profiles have occurred over time, in light of which modifications have been introduced into the epidemiological transition model. It was further developed by Olshansky and Ault (1986) and Olshansky, Carnes, Rogers and Smith (1998) with the addition of two further stages (4 and 5): 4. 'delayed degenerative diseases', a stage characterized by a decreasing rate of growth in deaths from degenerative diseases while these continue to dominate in overall mortality; 5. 'emerging infectious diseases', a stage which sees the recurrence of forgotten infectious and parasitic diseases. This occurs partly in the non-immunized population across various age groups, and partly among those who are vulnerable due to old age (Figure 9.2). The onset of this stage is accelerated by the processes of globalization.

A 'hybridic' stage has also been proposed (Rogers and Hackenberg 1987) in which mortality is connected with pathological behaviour, relating, in particular, to deaths resulting from traffic accidents, suicides, homicides, alcohol-related illnesses such as cirrhosis and HIV/AIDS. Such causes of death are termed social pathologies and it is argued that these became a significant factor in mortality as early as the 1970s, for example, in the United States (appropriately also the birthplace of the term hybridic stage).

Omran also introduced modifications into his theory, adding his own, similar to Olshansky, phases four and five to the classical model (1998).

It is now generally held by epidemiological transition theorists that developed countries have already entered stage five of the transition, but low- and medium-income countries, due to their financial and social limitations, are between stages two and three and have to face the challenge of fighting the double burden of infectious and non-infectious diseases. Countries of the former Soviet bloc are currently transitioning from stage three to stage four, or are already in the latter.

Omran (1971) emphasized that patterns of illness vary depending on geographical location and time period, and only marginally referenced the changing age structure of populations due to decreasing mortality. In general, however, he treated population as an indivisible whole, without age categorization. His main propositions were as follows:

1   the mortality rate is the most important factor in the dynamics of population size;
2   during epidemiological transition, changes in the mortality rate and the structure of mortality by cause of death occur in long cycles;
3   during epidemiological transition, the greatest changes in the changes in the mortality rate and the structure of mortality by cause of death occur among children and young women;
4   changes in a population's state of health and morbidity structure are closely linked to demographic and socio-economic shifts arising out of modernization;
5   on the basis variations between countries with regard to wealth, the rate of population change and the causes and effects of the latter, three basic models of epidemiological transition can be identified: the classical, Western model (England, Wales and Sweden); the accelerated model (Japan); and the contemporary, delayed model (Chile and Ceylon).

Models of epidemiological transition

1   The classical model describes a steady change from high mortality (death rate above 30 per cent) and high reproduction (birth rate above 40 per cent) to low mortality and low fertility (10 per cent and 20 per cent respectively), which occurred in conjunction with modernization processes experienced by most countries of Western Europe. After the 'pestilence and famine' stage

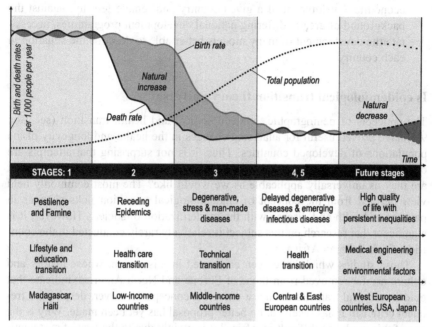

STAGES: 1	2	3	4, 5	Future stages
Pestilence and Famine	Receding Epidemics	Degenerative, stress & man-made diseases	Delayed degenerative diseases & emerging infectious diseases	High quality of life with persistent inequalities
Lifestyle and education transition	Health care transition	Technical transition	Health transition	Medical engineering & environmental factors
Madagascar, Haiti	Low-income countries	Middle-income countries	Central & East European countries	West European countries, USA, Japan

Source: Omran (1998); Olshansky and Ault (1986); Olshansky, Carnes, Rogers and Smith (1998).

Figure 9.2 The stages of the epidemiological transition.

in the pre-industrial period and at the beginning industrialization, mortality went into decline, but this only became clearly noticeable at the turn of the nineteenth and twentieth centuries, when there was an equally clear decrease in the female fertility rate. Social and economic changes in the populations of Western Europe were the undoubted cause of these declines in reproduction and mortality.

2 The accelerated model, which applies mainly to Japan, describes changes in mortality characteristic of the two successive stages: 'pestilence and famine' and 'receding epidemics'. These stages unfold in a similar fashion to the Classical model, but later in time, and stage transitions take place much more rapidly.

3 The contemporary (delayed) model applies to most developing countries. A small decline in mortality was seen in some of these countries as early as the beginning of the twentieth century, but a marked change occurred only after the Second World War. Thanks to international aid and Western medicine (including new drugs), mortality fell in a relatively short period of time (including deaths from prevalent infectious and parasitic diseases specific to given regions), while female fertility remained unchanged resulting in a process of demographic explosion. Nevertheless, the mortality rate for infants and children, and women of childbearing age remained extremely

high in these countries. The large variations between the level of socio-economic development of a given country and female fertility, against the background of greatly differing national development programmes, necessitate the development of many models better able to describe the situation in each country.

Is epidemiological transition theory universal?

The concepts of demographic, epidemiological and health transition (see Box 9.1) have proved useful for analysing changes in the health and longevity of the populations of developed countries. Thus, it is not surprising that attempts are continually made to also use these concepts in developing countries as well. But are they as universally applicable as we would like? The most commonly held view derived from the concept of epidemiological transition links changes in patterns of health and disease with the modernization of society. However, it is significant that research on this subject is relatively rarely conducted in the countries of sub-Saharan Africa.

Those studies which have been conducted in this region, whose history and culture are far removed from those of Europe and North America, indicate that epidemiological transition is not a universal concept. However, despite the frequent criticism it has received, no better proposal has yet been made. Why is the use of this model so difficult in Africa? It is mainly due to the lack of systematically maintained medical statistics, scarce data for morbidity and in too many cases the absence of clearly defined causes of death, all of which drastically diminish the information available as to public health in the region. It is easier in this situation to assess the epidemiological transition stage of a given country, by using risk factors instead of causes of death as the statistical material to feed into the demographic and epidemiological transition models. The UN most commonly estimates adult mortality by matching indices of child survival to model age patterns of mortality. Thus, the epidemiological transition stage assigned for any particular sub-Saharan country is also an estimate.

Another 'anomaly' with respect to epidemiological transition theory is the 'double burden of disease', a phenomenon which has been become evident in developing countries over the last three decades. In many of these countries, given the prevalence of malaria, TB and the HIV/AIDS pandemic, infectious and parasitic diseases undoubtedly continue to be a serious problem, but ever more frequently these are now accompanied by emerging non-infectious diseases linked to poor diet and unhealthy lifestyles.

Although the double burden of disease is particularly visible in countries with rising levels of income, it is also common in places where poverty and chronic hunger affect children pre-birth predisposing them to an increased risk of the early onset of non-communicable disease in later life. In low- and middle-income countries, the incidence of both infectious and non-infectious diseases is strongly dependent on socio-economic status, food security, the ability of individuals, families and communities to obtain affordable food throughout the year

---

Box 9.1 Health transition theory

In 1991, Frenk *et al.* (1991b) introduced the concept of health transition to link the changing stages in mortality with those in state of health and health care. They argue that state of health is determined by many complex, multi-level interactions at each stage of transition.

This concept was interestingly developed in a paper by Vallin and Meslé (2004). The authors formulate the proposition that epidemiological transition is the first stage in a global health transition process, which not only includes changes in the state of health of various populations, but also the interrelationship between a population, its health situation and the social response to changes in the state of public health. The tendency of the health transition depends on epidemiological factors and their development over time, as well as on behavioural, cultural and social factors, including eating habits, lifestyle, level of education and the sense of control over life and health.

According to Caldwell (Caldwell and Caldwell 1991; Caldwell 1992), there is a greater and more complex dependency between state of health and cultural, social and lifestyle factors (especially with regard to child mortality) than is commonly believed. Particularly strong disproportions can be seen in relation to the impact of the education received by parents, and mothers in particular. In developing countries today, greater education among mothers translates into growing numbers of vaccinations received by children, more attention to cleanliness at home and an increased understanding of the need for professional care in the event of illness. It is not so much the number of years spent at school which matters, as the mere fact of attending school, even if just briefly. In addition, Caldwell (1986) provided a historical perspective on the paths by which a decline in mortality in poor countries can be achieved. These are: 1. The availability of a modern health service; 2. Education, especially of mothers; 3. Family planning; 4. The characteristics of the society, including: a) the level of women's independence, b) egalitarianism, or c) radical traditionalism; and 5. The communist path to development. All these methods may be associated with an increase in the standard of living of a country's population, albeit in many cases loosely. In addition, communist development led also to increased mortality due to hunger, for example, in Ukraine and the south-western part of the former USSR.

---

and maintain access to sources of drinking water, as well as the skilful use of food preservation methods (Boutayeb 2006; Amuna and Zotor 2008).

Africa

Africa is currently the only continent where the number of deaths from communicable diseases exceeds the number of deaths from civilization diseases[3] and injuries. The traditional challenges for public health in Africa are HIV/AIDS, tuberculosis, malaria, pneumonia and diarrhoea. However, over the past three decades, there has been a sharp increase in the incidence of coronary heart disease, strokes, hypertension, diabetes, asthma, chronic liver and kidney disease. In sub-Saharan Africa, prevalence rates of type 2 diabetes and cardio-vascular disease (CVD) rose tenfold between 1988 and 2008 (Amuna and Zator 2008). The double burden of disease is overwhelming the capacity of the already poorly funded health services of Africa.

Barthelemy Defo (2014) provides an analysis of regional and national patterns of mortality, fertility, population growth and causes of death between 1950 and 2010 for all African countries based on the existing literature and available statistical data from such international organizations as WHO and UNICEF. Then, using a few relatively simple methods of analysing changes in the quantitative indicators, he examines how the concepts of demographic, epidemiological and health transitions correspond with the above-mentioned patterns over the sixty years under review. He concludes that the theoretical and empirical evidence indicates that the transition theories have a number of limitations in relation to Africa. For this reason, he proposes a new conceptual framework for the continent, which he calls the 'eco-epidemiological life course framework'. Defo's statistical analysis gives us a picture of Africa as a continent of uncertainty and disasters, with its trends in health, morbidity and mortality characterized by discontinuities and sudden changes. This portrayal does not fit within the demographic, epidemiological and health transition frameworks in every scale, from regional to national and transnational.

In the second half of the twentieth century, there was a very clear fall in African mortality rates. Unfortunately, in the twenty-first century, this is no longer the norm, as Pison notes (Plewes and Kinsella 2012: 6–7). In many countries, not only has no progress been made in this area, there has even been a noticeable deterioration. And at the same time, the classical epidemiological transition concept began to increasingly diverge from African reality.

The analysis by Defo (2014) shows in detail that in the sixty years to 2010 Africa as a whole and sub-Saharan Africa in particular remained the poorest region in the world, without any noticeable improvement in the health and longevity of its population. The following findings presented by Defo seem of particular significance:

1   there is a clear variation in trends in health care and mortality patterns as well as fertility and life expectancy between African countries and regions;

2   from 1950 to 1990, there was a dramatic decrease in mortality especially of infants and an increase in life expectancy across the continent;

3   since 1990, adult mortality rates have been increasing, which is primarily attributable to HIV/AIDS and accompanying illnesses. This phenomenon has ended the trend of rising life expectancy in Africa, and in several countries, those with the highest incidence of HIV/AIDS, it has reversed the general rule of longer life expectancy for women;

4   after 1990, in the sub-Saharan countries, wars have been a key factor in destroying the favourable mortality trend for children under five years old, especially in Central and Eastern Africa.

In research concerning Africa, nothing is obvious. Although most countries of eastern and southern Africa are experiencing immense adult over-mortality, which is clearly correlated with an earlier increase in HIV infections, in East Africa, by happy contrast, adult mortality began to decline even before the expansion of antiretroviral therapy programmes in 2005. Even more surprising is the lack of any consistent improvement in adult survival rates in African countries not impacted by major HIV epidemics.

Pison draws attention (Plewes and Kinsella 2012: 6–7) to the interesting case of Senegal, a low AIDS prevalence country, where the mortality of children fell in the 1970s and 1980s, then plateaued in the 1990s, and started falling again at the beginning of the twenty-first century. The initial two decades of success resulted from a vaccination programme, while mortality cast its shadow over the 1990s due to recurrent malaria and the lack of further progress with vaccinations. Only a resurgence of efforts to reduce the incidence of malaria (mosquito nets, new diagnostic tests and medicines), as well as increased vaccination coverage for other infectious childhood diseases, allowed many young lives to be saved in the twenty-first century. Furthermore, these successes were accompanied by other positive side effects such as a simultaneous reduction in mortality from diseases other than those against which the eradication efforts were specifically directed. This was probably because mosquito nets not only protect against malaria, but also against other diseases for which mosquitoes are vectors. It might be assumed that the improved socio-economic conditions noted in Senegal in the new millennium would also have played a role but for the fact that effects in areas such as mortality tend only to appear in the longer term.

Because chronic diseases develop over years or even decades, it is possible, as does the UN, to predict the incidence of such diseases on the basis of the occurrence of risk factors. Thus in Africa, bad diet and the tendency to obesity in certain sections of the population point to a possible increase in diabetes, CVD and some types of cancer. On the other hand, however, generally low levels of blood sugar and cholesterol found in Africans, together with smoking being less common than on other continents, means that estimates of morbidity and mortality rates for civilization diseases are also at a lower level. Although this is the overall picture, there are differences between countries, regions, and even between the incidence of health risk factors in rural and urban areas. Ezzati (Plewes and Kinsella 2012: 8), based on GBD (Global Burden of Diseases), showed that:

1   everywhere in the world, increases in body weight and glycemia have been observed, except in a few regions including sub-Saharan Africa (for the male population);
2   average blood pressures are decreasing in developed countries, remaining relatively stable in East Asia and rising in sub-Saharan Africa[4];
3   in Europe and North America, there are declining serum total cholesterol levels, while in South and East Asia they are rising, and in Africa we can see a mix of these trends. According to Vorster (2002), the available data in

South Africa suggests that black Africans may be resistant to ischemic heart disease (IHD) due to their serum lipid profiles (low cholesterol and high ratios of high-density lipoprotein cholesterol), low homocysteine values and genetically determined low homocysteine values. However, increased animal fat and protein intake among affluent black South Africans, who can afford a Western diet, is leading to high BMI growth in men and women, and raised serum total cholesterol levels. It is assumed that such risk factors, occurring at different stages of urbanization, will increase the risk of future non-communicable diseases (NCDs), including IHDs and strokes (Table 9.1).

Based on research published on PubMed[5] relating to selected local communities and countries, Dalal *et al.* (2011) determined the variance in the incidence of NCDs, and associated risk factors, between countries and urban versus rural locations. The incidence of strokes ranged from 0.07 to 0.3 per cent, diabetes from 0 to 16 per cent, hypertension 6 to 48 per cent, obesity 0.4 to 43 per cent and smoking 0.4 to 71 per cent. Hypertension was found to be equally common in men and women, obesity was more common among women and males were more likely to be smokers (Dalal *et al.* 2011). A concept of epidemiological transition, which does not take into account the distinction between rural and urban populations, is of little use in assessing a country's health situation. Certainly, the help it can give to African governments in health policy planning is very limited.

One of the interesting studies in Africa of the demographic transition on a local scale was conducted in Accra, the capital of Ghana, which was in the past a rapidly developing colonial city and already of key interest in the literature on culture, development and health in Ghana (Agyei-Mensah and de-Graft Aikins

Table 9.1 Chronic non-communicable disease as a new epidemic in Africa: focus on The Gambia

	Cardiovascular diseases	Diabetes	Asthma	Others	Total
2008					
Morbidity	47,258	2,153	8,934	422	58,767
Admission	1,454	182	432	135	2,203
Mortality	118	11	13	25	167
2010					
Morbidity	54,225	3,115	10,174	323	67,837
Admission	1,458	176	450	128	2,212
Mortality	72	18	18	25	133
2011					
Morbidity	57,208	3,232	9,621	322	70,383
Admission	1,532	203	501	184	2,420
Mortality	145	19	17	25	206

Source: Omoleke (2013).

2010). The study focuses on Accra's transition from a colonial to a global city and how demographic, economic, socio-economic and geopolitical changes have affected the city's health. The epidemiological transition in Accra is described as a protracted polarized model, a direct reference to the double burden of infectious and chronic disease. The prosperous section of the city's population is more likely to suffer from civilization diseases, whereas those who live in poverty suffer from infectious diseases or from both groups of diseases. According to Agyei-Mensah and de-Graft Aikins (2010), this situation arises from a growing degree of urbanization (accompanied by urban poverty) and globalization.

Asia

The double burden of disease is also pervasive in Asia. And, as in Africa, there are marked differences in health between countries, regions and local areas. The main common risk factors are excessive calorie consumption, low hygiene, unhealthy lifestyles and genetic predisposition.

According to Tollman (Plewes and Kinsella 2012: 17) in Asia, unlike in Africa, NCDs are currently the leading cause of adult deaths in almost every country. However, the types of NCDs prevalent in each country vary considerably. For example, heart attacks are the predominant cause of death in India and strokes in China. Indians suffer more frequently from oral and lung cancer, whereas for the Chinese lung cancer is more common (it is the leading cause of death among Chinese adults).

The increasing prevalence of NCDs in Asia is a consequence of various factors, central among which is the dramatic fall in fertility over the last few decades in many Asian nations. This drop, along with a decrease in childhood mortality, has caused an age shift and changes in the causes of death, which is precisely what we describe as epidemiological transition (WHO 2011).

China

China's health situation has improved significantly over the past sixty years and more, in line with the classical model of epidemiological transition. As recently as the early 1970s, the country was characterized by high rates of infectious diseases and high mortality among the poor peasantry. But the following decades saw a decrease in the incidence of infectious diseases, and growth in morbidity and mortality associated with the ageing of an increasingly city-based population. This process has given rise to new health problems: chronic and degenerative diseases. Recent years have clearly signalled China's entry into the next stage of the epidemiological transition, characterized by an increase in the average life expectancy and the prevalence of civilization diseases in conjunction with the emergence and re-emergence of infectious diseases (Cook, Trevor and Dummer 2004).

Today in China there is a new set of health problems, including conditions related to smoking, hypertension and environmental pollution as well as an increase in HIV/AIDS infections. In spite of the tremendous acceleration of the country's economic development, the challenge in China is to maintain what has already been achieved in the field of health while at the same time addressing the new problems associated with rapid urbanization, widening social inequalities, the spread of HIV/AIDS and the emergence of other new infectious diseases (Cook *et al.* 2004).

Bangladesh

Like many other developing countries, Bangladesh finds itself between the second and third phases of the epidemiological transition, and therefore has many health burdens associated with both malnutrition and obesity, continuing high morbidity from infectious diseases and the increasing incidence of NCDs. Infectious diseases which are prevalent in Bangladesh include HIV/AIDS, tuberculosis, malaria, leprosy, pneumonia, diarrhoeal diseases, tuberculosis, measles and vector-borne diseases, like dengue, visceral leishmaniasis (kala azar) and filariasis. Within this group, an important and relatively new problem is posed by drug-resistant infectious diseases and re-emerging diseases like Chikungunya (Mahmood, Ali and Islam 2013; Mascie-Taylor 2012).

South America

The current health situation in many countries of South America also does not conform to the classical model of epidemiological transition. This is due to the resurgence of old infectious diseases (malaria and dengue) and variations in the proportions of infectious and non-infectious diseases.

Studies on South America have long noted the continent's heterogeneous health profile in which different countries are in various stages of epidemiological transition.

> in most of them the transition experience is unlike that of the developed countries and is distinguished by: (a) a simultaneous high incidence of diseases from both the pre- and post-transitional stages; (b) a resurgence of some infectious diseases that had previously been under control; (c) a lack of resolution of the transition process, so that the countries appear to be caught in a state of mixed morbidity; (d) a peculiar epidemiological polarization, not only between countries but also in the different geographical areas and between the various social classes of a single country. This experience is called a 'prolonged polarized model'.
>
> (Frenk at al. 1991a: 485)

Mexico

Since 1990, the mortality of children and mothers in Mexico has decreased, but the mortality of adults has increased. The main risk factors to health in the

1990s, such as undernutrition, have today been replaced by rising plasma glucose and body mass index levels. Adult mortality is increasing due to chronic kidney disease, diabetes, cirrhosis of the liver, and especially since 2000, as a result of violence, especially among men. The incidence of diarrhoeal diseases and protein-energy malnutrition has fallen significantly, while chronic kidney disease and breast cancer in women have begun to increase sharply. In 2013, the main causes of 'disability-adjusted life years' (DALYs)[6] were diabetes, ischaemic heart disease, chronic kidney disease, low back and neck pain, and depressive disorders, the last three of which appeared on the list only after 1990, indicating an epidemiological transition from the prevalence of infectious diseases to that of civilization diseases. However, although differences in public health between the states of Mexico have diminished over time, they have increased at the local level (Gómez-Dantés *et al.* 2016).

Chile

Chile's economy has been growing strongly since the mid-1990s, but the number of infectious diseases (e.g. TB, gastrointestinal diseases and STDs) remains high and this places the country at a transitional stage of development. However, there has been a decrease in the incidence of diseases easily preventable through immunization.

Between 1970 and 1992, there was a decline in the fertility rate from 3.4 to 2.6, with a significant reduction in overall infant mortality, resulting in an increase in life expectancy of eight years for men and nine for women. This is turn led to alterations in the age structure of the population and age-correlated changes in the predominant causes of morbidity and mortality. Reflecting these changes, NCD death rates increased from 53.7 per cent of all deaths in 1970 to 74.9 per cent in 1991 (Albala and Vio 1995).

Brazil

An analysis of changes in the pattern of morbidity and mortality in Brazil since the 1940s, carried out by Prata (1992), led him to conclude that the country was in the process of an 'incomplete epidemiological transition', as, in addition to the large proportion of deaths due to chronic degenerative diseases, there had been a rise in mortality due to the resurgence of infectious diseases which were thought to have been eradicated. In modern Brazil, infectious diseases predominate among the poor, while the wealthy suffer from diseases of civilization. In many of Brazil's communities, heart disease, cancer and strokes cannot be categorized as degenerative diseases as they do not correlate with a growing median age of those affected. Rather, they are the result of an inhospitable natural environment and socio-cultural factors which are conducive to the spread of infectious and civilization diseases alike.

The inhabitants of the Tupe Sustainable Development Reserve (RDS Tupé), west of the city of Manaus in the Amazon rainforest also suffer from two major groups of diseases: the one parasitic and infectious, and the other chronic and degenerative. In order to discover what lay behind RDS Tupé's diverse morbidity and mortality profile, a wide range of factors – political, economic, lifestyle, cultural and environmental – were examined (Mariosa, Dota, Gigliotti and dos Santos-Silva, 2015). It was concluded that the major single determiner in relation to infectious diseases was the Reserve's location on the boundaries of the metropolis and the damp Equatorial forest. Age and level of education were statistically significant only for degenerative diseases.

Central Europe

The processes of changes in health which have taken place in Central Europe are well described by the stages of the epidemiological and health transition. Thus, despite the delaying effect brought about by the period of socialist economy, the countries in this part of Europe are in line with the classical model of epidemiological transition.

Poland

Analysis of the historical data indicates that for Poland the processes of passing through the first stages of the transition were similar to the experience of the countries of Western Europe, but with time lags. The levels of mortality and life expectancy in ethnically Polish territories at the end of the nineteenth century indicate that the 'pestilence and famine' era was drawing to a close, and the second phase of epidemiological transition was beginning. The 1920s and 1930s may have been when Poland transitioned out of the second phase of epidemiological transition (a progressive decline in the incidence of infectious diseases). In the 1950s, overall mortality rates, including infant mortality, and average life expectancy were at the level found in the third phase of the epidemiological transition. In the mid-1970s, they reached the upper limit of values for this stage. In the 1990s for women, and from 2000 at the latest for men, a new stage of health was reached in Poland, indicating entry into the fourth phase of the epidemiological transition. The very low and declining contribution of infectious diseases to Poland's mortality structure indicates that there is no resurgence of infectious diseases as causes of death in the country's population. A life expectancy of seventy years and over which characterizes the completion of the third phase of epidemiological transition was reached by Polish women in the early 1960s, while for men this only occurred in 2001 (Wróblewska 2009).

Civilization diseases, such as CVD and cancer, have been the leading causes of death in Poland for many years and account for more than 70 per cent of the total number of deaths in men and 80 per cent of deaths among women. Reductions in CVD mortality, both for men and women, occurred in the early 1990s. High CVD mortality rates, characteristic of the elderly, have fallen significantly, which in recent years has been reflected in a significant increase in life expectancy, first among women and then among men. To a lesser extent, and with time lags, there has been a decline in mortality due to cancer (Wróblewska 2009).

The Omran model with modifications has proven to be a valid and useful tool with which to analyse changing patterns of illness in Poland over historical time.

The Western perspective on the key determinant of health

Regardless of the breadth and depth of discourse regarding health, the proverbial 'man in the street' remains convinced that one of the best ways to prolong life is money, which can buy the ability to live with dignity thanks to the numerous medical and technical possibilities the twenty-first century offers. In the literature, living with dignity is often defined by the bipolar term 'quality of life' (from poor to very good). Determinants of health, in the broadest sense (not just those related to access to money), have been investigated by Nobel Prize-winning economists (Nobel Prize n.d.) such as Kenneth J. Arrow (Nobel Laureate in 1972, who 'contributed to the general economic equilibrium theory and welfare theory'), Amartya Sen (in 1998, 'for his contributions to welfare economics'), Joseph E. Stiglitz, George Akerlof and Michael Spence (in 2001, 'for their analyses of markets with asymmetric information'), Angus Deaton (in 2015, 'for his analysis of consumption, poverty, and welfare'). Another recipient of the Nobel Prize in economics – Alvin E. Roth (in 2012, with Lloyd S. Shapley) – is one of the founders and designers of the New England Program for Kidney Exchange.

According to Michał Rutkowski (2007), one of the directors of the World Bank in Washington (Rutkowski n.d.), the fatalistic perspective that the level of a country's health is contingent on its wealth and automatically grows in line with increasing prosperity has dominated for too long among economists. Perhaps this view has been so persistent due to the difficult circumstances which exist in less economically developed countries? In reality, however, neither the level of wealth, nor the GDP per capita growth rate have had as much influence on epidemiological transition as originally thought.

The fact is that the science of health and its determinants is dominated by the Western point of view. But today, it seems that some reflection is beginning to stir on this subject, also in the West. The Royal Geographical Society (with the Institute of British Geographers) devoted its Annual Conference in 2017 to the subject: *Decolonizing Geographical Knowledges: Opening Geography Out to the World* (RGS-IBT 2017).

Nevertheless, it is thanks to the work of Western economists that we know that in developed countries, the health and well-being of citizens (as measured by the mortality rate) no longer correlate with GDP growth per capita but rather the degree of social inequality. In the past, in the developed world, a rise in living standards was a direct cause of declining mortality. At present in these countries, health depends more heavily on income distribution than on economic growth. The primacy of material limitations has given way to the primacy of social constraints as factors conditioning the quality of human life. This change does not only have implications for health. Health is merely the effect of social and economic change, a measure of well-being (Wilkinson 1994).[7]

Inequality of opportunity, defined as differences in the prospects for social advancement, also has a significant impact on health. Hope for social advancement is highly correlated with good health and healthy behaviour on the part of individuals. And with reduced opportunities the situation is reversed, and there is lower motivation to invest in future health (Venkataramani *et al.* 2016). Perhaps inequality of opportunity is the real factor which models the health of a population (Katikireddi 2016). It can be assumed that further measures to expand economic opportunities and life chances (not just financial benefits in themselves) may have profound and positive effects on the health of local communities in many developed countries. And perhaps in the longer term also in underdeveloped countries.

Migration and disease

Migration is described as a window to the world. There are economic, demographic, cultural and social transitions connected with various forms of migration, which affect the speed and nature of epidemiological transition. Just the migratory journey alone, of course, forces immigrants to change their physical environment, and this in itself can be potentially beneficial or detrimental (Plewes and Kinsella 2012: 10).

The mass migrations of the era of globalization are blamed for creating exceptionally favourable conditions for epidemic and pandemic outbreaks of dangerous infectious diseases, including our 'old friends' cholera and tuberculosis (TB), or new ones such as further mutations of the influenza virus (Figure 9.3).

In Europe, renewed TB elimination is a key public health task. With the rapid influx of immigrants from underdeveloped countries, reports have appeared in the media warning of the danger of a resurgence of TB. The UK, Sweden, the Netherlands and Norway (Hargreaves, Nellums and Friedland 2016) have all reported an increase of over 70 per cent in new incidences of the disease.

Screening at border crossing points indicates that migrants from countries with a high incidence of TB who have been treated for the disease represent a

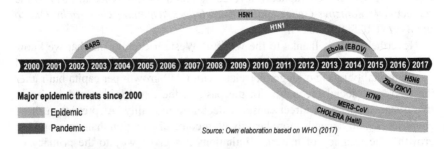

Figure 9.3 Global epidemics in the twenty-first century.

negligible transmission risk. However, they themselves are more likely to contract TB, if they are in contact an infected person. This threat could potentially be avoided if all travellers entering the EU were routinely checked in order to identify those with latent TB infections. In TB screening among migrants crossing into the UK between 2006 and 2012, the incidence of bacteriologically confirmed pulmonary tuberculosis was estimated at 49 per 100,000 (Aldridge *et al*. 2016).

The globalization of trade, leading to the long-distance transport of goods, animals and food products, provides favourable conditions for the worldwide spread of drug-resistant pathogens. An example is the proliferation of Salmonella spp., which was investigated by Polish scholars using current data on resistance among Salmonella spp. isolates of food origin from countries in different regions of the world. Their findings demonstrate that the 'global transport' of drug-resistant bacteria is a real possibility. Nobody in the world, even someone who does not travel, is free from the risk of contracting resistant strains of bacteria. They can be delivered to us in any product circulated in international trade (Mąka and Popowska 2016).

Conclusion

Regardless of where we live, there is one thing we can be certain of in life and that is change. At the time when the epidemiological transition theory and the associated concept of health transition were developed, they were a source of hope for they mapped successive changes for the better. They held out the promise that the health of individuals and whole populations would continue to improve and thus that humanity would enjoy greater longevity. This prospect is no doubt the reason for the great popularity of Omran's theory which enjoyed surging interest especially in the 1990s, just some twenty years after its publication. His theory elegantly supports the assumptions of globalization. Subsequent modifications to the theory have gradually extinguished the possibilist enthusiasm expressed in the belief that humans are able to control the influence of the natural environment and its threats to our health. Meanwhile, recent years have seen the growing popularity of the nihilist view that political stability, scientific and technical development, and legal order may be crucial to our future health. At the same time, every successive epidemic of the twenty-first century has prompted repeated discussion skirting on the edges of geographic determinism, according to which even well-developed countries located in zones with environmental conditions conducive to the development of infectious and parasitic diseases, will find difficult it to protect their populations. All in all, after years of hope, there has come a time of doubt in the possibility of positive health changes on a global scale. Health inequalities within countries and regions are rather causing concerns about the global migration of pathogens than the joyful expectation of global medical progress.

According to Polanco (2016), new factors will continually emerge to contribute to the outbreak of further epidemics, and these will inevitably involve a significant number of human deaths. It is therefore necessary to create a robust nanotechnology-based early warning system allowing the real-time collection of biometric data from travellers (e.g. at borders) in order to immediately signal imminent health threats. This would help minimize the impact of future epidemics, but are we ready to accept yet another form of surveillance?

Notes

1 Morbidity – the prevalence of a disease in a particular percentage of the population; the number of cases of a particular disease per unit of population (Farlex Partner Medical Dictionary 2012).

  Mortality – the death rate, which reflects the number of deaths per unit of population in any specific region, age group, disease or other classification, usually expressed as deaths per 1,000, 10,000 or 100,000 (Farlex Partner Medical Dictionary 2012).

2 Degenerative diseases – diseases resulting from structural damage to tissues or organs (e.g. atherosclerosis, Alzheimer's, Parkinson's).

3 Civilization diseases are defined as globally occurring, widespread diseases, whose emergence or proliferation is linked to the advancement of contemporary civilization. Thus, such conditions (e.g. heart and circulatory diseases, diabetes, cancer, depression, etc.) are also known as 'twenty-first-century diseases'.

4 Over the past forty years, there has been a noticeable change in the incidence of high blood pressure among people in different parts of the world. Significant growth has occurred among South Asian and sub-Saharan people, while among Central and Eastern Europeans the level has remained stable though high (Matheson *et al.* 2013).

5 PubMed is a free resource that is developed and maintained by the National Center for Biotechnology Information (NCBI), at the U.S. National Library of Medicine (NLM), located at the National Institutes of Health (NIH), www.ncbi.nlm.nih.gov/pubmed/.

6 DALY can be thought of as one lost year of 'healthy' life.

7 Therefore, we should consider whether non-sustainable economic growth is worth the risk to the natural and human environment. A relevant example here is China, which has definitely emerged from poverty as a country (although not all its citizens have benefitted) and has achieved spectacular economic success, but the environmental and health cost has been high.

References

Agyei-Mensah, S. and de-Graft Aikins, A. (2010) Epidemiological Transition and the Double Burden of Disease in Accra, Ghana. *Journal of Urban Health: Bulletin of the New York Academy of Medicine* 87(5): 879–897.

Abala, C. and Vio, F. (1995) Epidemiological Transition in Latin America: The Case of Chile. *Public Health* 109: 431–442.

Aldridge, R.W., Zenner, D., White, P.J., Williamson, E.J., Muzyamba, M.C., Dhavan, P., Mosca, D., Thomas, H.L., Lalor, M.K., Abubakar, I. and Hayward, A.C. (2016) Tuberculosis in Migrants Moving from High-Incidence To Low-Incidence Countries: A Population-Based Cohort Study of 519 955 Migrants Screened Before Entry to England, Wales, and Northern Ireland. *The Lancet*, 388(10059): 2510–2518, Retrieved from www.thelancet.com/pdfs/journals/lancet/PIIS0140-6736(16)31008-X.pdf.

Amuna, P. and Zotor, F.B. (2008) Epidemiological and Nutrition Transition in Developing Countries: Impact on Human Health and Development: The Epidemiological and Nutrition Transition in Developing Countries: Evolving Trends and their Impact in

Public Health and Human Development. *Proceedings of the Nutrition Society* 67(1): 82–90. Retrieved from www.cambridge.org/core/journals/proceedings-of-the-nutrition-society/article/div-classtitleepidemiological-and-nutrition-transition-in-developing-countries-impact-on-human-health-and-developmentdiv/0764E5FE03E5C92D39CAA 6184B5998EE.

Bhutta, Z.A. (2016) Mapping the Geography of Child Mortality: A Key Step in Addressing Disparities. *The Lancet* 4(12): e877–e878. Retrieved from www.thelancet.com/pdfs/journals/langlo/PIIS2214-109X(16)30264-9.pdf.

Boutayeb, A. (2006) The Double Burden of Communicable and Non-Communicable Diseases in Developing Countries. *Transactions of the Royal Society of Tropical Medicine and Hygiene* 100(3): 191–199.

Burch, L.S., Smith, C.J., Anderson, J., Sherr, L., Rodger, A.J., O'Connell, R., Geretti, A.-M., Gilson, R., Fisher, M., Elford, J., Jones, M., Collins, S., Azad, Y., Phillips, A.N., Speakman, A., Johnson, M.A. and Lampe, F.C. (2016) Socioeconomic Status and Treatment Outcomes for Individuals with HIV on Antiretroviral Treatment in the UK: cross-sectional and longitudinal analyses. *The Lancet Public Health*: e26–36. Retrieved from www.thelancet.com/pdfs/journals/lanpub/PIIS2468-2667(16)30002-0.pdf.

Caldwell, J.C. (1986) Routes to Low Mortality in Poor Countries. *Population and Development Review* 12(2): 171–220.

Caldwell, J.C. (1992) Old and New Factor in Health Transition. *Health Transition Review* 2, supplement issue: 205–215.

Caldwell, J.C. (2001) Population Health in Transition. *Bulletin of the World Health Organization* 79(2): 159–170.

Caldwell, J.C. and Caldwell, P. (1991) What have we Learnt About the Cultural, Social and Behavioural Determinants of Health?: From Selected Readings to the First Health Transition Workshop. *Health Transition Review* 1(1): 1–3.

Cook, J.G., Trevor, J.B., and Dummer, T.J.B. (2004) Changing Health in China: Re-Evaluating the Epidemiological Transition Model. *Health Policy* 67: 329–343.

Dalal, S., Beunza, J.J., Volmink, J., Adebamowo, C., Bajunirwe, F., Njelekela, M., Mozaffarian, D., Fawzi, W., Willett, W., Adami, H.O. and Holmes, M.D. (2011) Non-Communicable Diseases in Sub-Saharan Africa: What We Know Now. *International Journal of Epidemiology* 40(4): 885–901.

Defo, B.K. (2014) Demographic, Epidemiological, and Health Transitions: Are They Relevant to Population Health Patterns in Africa?. *Global Health Action* 7(1), Retrieved from www.tandfonline.com/doi/full/10.3402/gha.v7.22443?scroll=top&needAccess=true.

Evans, R.G., Barer, M.L. and Marmor, T.R. (1994) *Why are Some People Healthy and Others Not?: The Determinants of Health Populations, Social Institutions and Social Change Series*. New York: Transaction Publisher.

Farlex Partner Medical Dictionary (2012) Retrieved from http://medical-dictionary. thefreedictionary.com/morbidity; retrieved from http://medical-dictionary.thefree dictionary.com/mortality.

Frenk, J., Bobadilla, J.L., Stern, C., Frejka, T. and Lozano, R. (1991a) Elements for a Theory of the Health Transition. *Health Transition Review* 1(1): 21–38.

Frenk, J., Frejka, T., Bobadilla, J.L., Stern, C., Lozano, R., Sepúlveda, J. and José, M. (1991b) The Epidemiologic Transition in Latin America. *Boletin de la Oficina Sanitaria Panamericana: Pan American Sanitary Bureau* 111(6): 485–496.

Gómez-Dantés H. at al. (2016) Dissonant Health Transition in the States of Mexico, 1990–2013: A Systematic Analysis for the Global Burden of Disease Study 2013. *The Lancet* 388: 2386–402.

Hargreaves, S., Nellums, L. and Friedland, J.S. (2016) Time to Rethink Approaches to Migrant Health Screening. *The Lancet* 388(10059): 2456–2457. Retrieved from www. thelancet.com/pdfs/journals/lancet/PIIS0140-6736(16)31703-2.pdf.

Horton, R. (2016) Offline: Planetary Health – Gains and Challenges. *The Lancet*, 388(10059): 2462. Retrieved from www.thelancet.com/pdfs/journals/lancet/PIIS0140-6736(16)32215-2.pdf.

Katikireddi, S.V. (2016) Economic Opportunity: A Determinant of Health?. *The Lancet Public Health* 1(1): e4–5. Retrieved from www.thelancet.com/pdfs/journals/lanpub/PIIS2468-2667(16)30004-4.pdf.

Laliotis, J., Ioannidis, J.P.A. and Stavropoulou, Ch. (2016) Total and Cause-Specific Mortality Before and After the Onset of the Greek Economic Crisis: An Interrupted Time-Series Analysis. *The Lancet Public Health* 1(2): e56–65. Retrieved from www. thelancet.com/pdfs/journals/lanpub/PIIS2468-2667(16)30018-4.pdf.

Link, B.G. and Phelan, J. (2002) McKeown and the Idea that Social Conditions are Fundamental Causes of Disease. *American Journal of Public Health* 92(5): 730–732. Retrieved from www.ncbi.nlm.nih.gov/pmc/articles/PMC1447154/.

McKee, M. and Stuckler, D. (2016) Health Effects of the Financial Crisis: Lessons from Greece. *The Lancet Public Health* 1(2): e56. Retrieved from www.thelancet.com/pdfs/journals/lanpub/PIIS2468-2667(16)30016-0.pdf.

McKeown, T., Record, R.G. and Turner, R.D. (1975) An Interpretation of the Decline of Mortality in England and Wales during the Twentieth Century. *Population Studies* 29: 391–422.

Mahmood, A.S., Ali, S. and Islam, R. (2013) Shifting from Infectious Diseases to Non-Communicable Diseases: A Double Burden of Diseases in Bangladesh. *Journal of Public Health and Epidemiology* 5(11): 424–434. Retrieved from http://citeseerx.ist.psu.edu/viewdoc/download?doi=10.1.1.428.6156&rep=rep1&type=pdf.

Mąka, Ł. and Popowska, M. (2016) Antimicrobial Resistance of *Salmonella* spp. Isolated from food. *Roczniki Państwowego Zakładu Higieny* 67(4): 343–358. Retrieved from www.researchgate.net/publication/311473998_Antimicrobial_resistance_of_Salmonella_spp_isolated_from_food.

Mariosa, D.F., Dota, E.M., Gigliotti, M. and dos Santos-Silva, E. (2015) Environmental Vulnerability, Demographic and Epidemiological Transition on RDS Tupé, Manaus, Amazonas. *Hygeia* 11(20): 138–152.

Mascie-Taylor, N. (2012) Is Bangladesh Going Through an Epidemiological and Nutritional Transition?. *Collegium Antropologicum* 36(4): 1155–1159. Retrieved from www.ncbi.nlm.nih.gov/pubmed/23390805.

Matheson, G.O., Klügl, M., Engebretsen, L., Bendiksen, F., Blair, N.B., Börjesson, M., Budgett, R., Derman, W., Erdener, U., Ioannidis, J.P., Khan, K.M., Martinez, R., Van Mechelen, W., Mountjoy, M., Sallis, R.M., Schwellnus, M., Shultz, R., Soligard, T., Steffen, K., Sundberg, C.J., Weiler, R. and Ljungqvist, A. (2013) Prevention and Management of Non-Communicable Disease: The IOC Consensus Statement, Lausanne. *British Journal of Sports Medicine* 47: 1003–1011. Retrieved from http://bjsm.bmj.com/content/47/16/1003.full.pdf+html.

NCD Risk Factor Collaboration (2016) Worldwide Trends in Blood Pressure from 1975 to 2015: A Pooled Analysis of 1479 Population-Based Measurement Studies with 19·1 Million Participants. *The Lancet* 388(10059): 1–19. Retrieved from www.thelancet.com/pdfs/journals/lancet/PIIS0140-6736(16)31919-5.pdf.

Noble Prize (n.d.) Lists of Nobel Prizes and Laureates. Retrieved from www.nobelprize.org/nobel_prizes/lists/age.html.

Olshansky, S.J. and Ault, B. (1986) The Fourth Stage of the Epidemiologic Transition: The Age of Delayed Degenerative Diseases. *The Milbank Quarterly* 64(3): 355–391.

Olshansky, S.J., Carnes, B.A., Rogers, R.G. and Smith, L. (1998) Emerging Infection Diseases: The Fifth Stage of the Epidemiologic Transition?. *The World Health Statistics Quarterly* 51(4): 207–213.

Omoleke, S.A. (2013) Chronic Non-Communicable Disease as a New Epidemic in Africa: Focus on The Gambia. *The Pan African Medical Journal* 14: 87. Retrieved from www.panafrican-med-journal.com/content/article/14/87/full/.

Omran, A.R. (1971) The Epidemiologic Transition: A Theory of the Epidemiology of Population Change. *The Milbank Quarterly* (2005) 83(4): 731–57. Reprinted from *The Milbank Memorial Fund Quarterly* (1971) 49(4): 509–538.

Omran, A.R. (1998) The Epidemiologic Transition Theory Revisited Thirty Years Later. *The World Health Statistics Quarterly* 51(4): 99–119.

Plewes, T.J. and Kinsella, K. (2012) *The Continuing Epidemiological Transition in Sub-Saharan Africa: A Workshop Summary, Committee on Population Division of Behavioral and Social Sciences and Education.* Washington, DC: The National Academies Press.

Polanco, C. (2016) Models Oriented to Infectious Disease Outbreaks: Possible Futures, Reader Comments. *The Lancet Infectious Diseases* 16(12): 1305, Retrieved from www. thelancet.com/journals/laninf/article/PIIS1473-3099(16)30483-2/fulltext.

Prata, P.R. (1992) A transição epidemiológica no Brasil. *Cadernos de Saúde Pública* 8(2): 168–175. Retrieved from www.scielo.br/pdf/csp/v8n2/v8n2a08.pdf.

Regidor, F., Vallejo, F., Tapia Granados, J.A., Viciana-Fernández, F.J., de la Fuente, L. and Barrio, G. (2016) Mortality Decrease According to Socioeconomic Groups during the Economic Crisis in Spain: A Cohort Study of 36 Million People. *The Lancet*, 388(10060): 2642–2652.

RGS-IBT (2017) *Decolonizing Geographical Knowledges: Opening Geography out to the World.* RGS-IBG Annual International Conference 2017, 29 August–1 September 2017. Retrieved from www.rgs.org/WhatsOn/ConferencesAndSeminars/Annual+International+Conference/Conference+theme.htm.

Rogers, R. and Hackenberg, R. (1987) Extending Epidemiologic Transition Theory: A New Stage. *Social Biology* 34(3–4): 234–243.

Rutkowski, M. (n.d.) About Michal Rutkowski. World Bank. Retrieved from www. worldbank.org/en/about/people/m/michal-rutkowski.

Rutkowski, M. (2007) Przejście epidemiologiczne. *Gazeta Bankowa* 35(985). Retrieved from www.gazetabankowa.pl/pl/dokumenty/nr_35_2.

Schreter, S. (2002) Rethinking McKeown: The Relationship Between Public Health and Social Change. *America Journal of Public Health* 92(5): 722–5.

Smith, T.G. (2016) Is Behavioural Economics Ready to Save the World?. *The Lancet Diabetes & Endocrinology* 4(12): 982.

UNICEF Office of Research (2013) Warunki i jakość życia dzieci w krajach rozwiniętych. Analiza porównawcza, Innocenti Report Card 11, Florence.

Vallin, J. and Meslé, F. (2004) Convergences and Divergences in Mortality: A New Approach to Health Transition, Demographic Research, Special Collection 2(2). Retrieved from www.demographicresearch.org.

van der Sande, M.A.B., Ceesay, S.M., Milligan, P.J.M., Nyan, O.A., Banya, W.A.S., Prentice, A., McAdam, K.P.W.J. and Walraven, G.E.L. (2001) Obesity and Undernutrition and Cardiovascular Risk Factors in Rural and Urban Gambian Communities. *American Journal of Public Health* 91(10): 1641–1644. Retrieved from http://ajph. aphapublications.org/doi/full/10.2105/AJPH.91.10.1641.

Venkataramani, A.S., Brigell, R., O'Brien, R., Chatterjee, P., Kawachi, I. and Tsai, A.C. (2016) Economic Opportunity, Health Behaviours, and Health Outcomes in the USA: A Population-Based Cross-Sectional Study. *The Lancet Public Health* 1(1): e18–25. Retrieved from www.thelancet.com/pdfs/journals/lanpub/PIIS2468-2667(16)30005-6. pdf.

Vorster, H.H. (2002) The Emergence of Cardiovascular Disease During Urbanisation of Africans. *Public Health Nutrition* 5(1a): 239–243. Retrieved from www.cambridge. org/core/services/aop-cambridge-core/content/view/9AADD871BD754C757B94E226 BE66772C/S1368980002000332a.pdf/the-emergence-of-cardiovascular-disease-during-urbanisation-of-africans.pdf.

WHO (2011) Noncommunicable Diseases in the South-East Asia Region. Retrieved from www.searo.who.int/nepal/mediacentre/2011_non_communicable_diseases_in_the_ south_east_asia_region.pdf.

WHO (2017) Disease Outbreaks by Year. Retrieved from www.who.int/csr/don/archive/ year/en/.

Wilkinson, R.G. (1994) The Epidemiological Transition: From Material Scarcity to Social Disadvantage?. *Daedalus* 123(4): 61–77.

Wojtyniak, B. and Goryński, P. (eds) (2016) *Sytuacja zdrowotna ludności Polski i jej uwarunkowania*. Warsaw: Narodowy Instytut Zdrowia Publicznego-Państwowy Zakład Higieny.

Wróblewska, W. (2009) Teoria przejścia epidemiologicznego oraz fakty na przełomie wieków w Polsce. *Studia Demograficzne* 1(155): 110–159. Retrieved from http://sd. pan.pl/old/images/stories/pliki/Archiwum/2009_1_6_ww.pdf.

Index

Page numbers in **bold** denote figures and those in *italic* denote tables.

Printed in the United States
by Baker & Taylor Publisher Services